PROGRESS IN
FLYING
MACHINES

———

Octave Chanute

———

With a new Introduction by
Joshua Stoff

DOVER PUBLICATIONS, INC.
Mineola, New York

Bibliographical Note

This Dover edition, first published in 1997, is an unabridged repub-
lication of the work first published by the *American Engineer and
Railroad Journal*, New York, in 1894. The Introduction to the Dover
Edition was prepared specially for this edition.

Library of Congress Cataloging-in-Publication Data

Chanute, Octave, 1832–1910.
    Progress in flying machines / Octave Chanute ; introduction by
Joshua Stoff. — Dover ed.
        p.   cm.
    Originally published: New York : The American Engineer & Rail-
road Journal, 1894.
    Includes bibliographical references and index.
    ISBN 0-486-29981-3 (pbk.)
    1. Flying-machines—History.   2. Aeronautics—History.   I. Title.
TL670.5.C5   1997
629.13—DC21   97–38658
                                                                      CIP

Manufactured in the United States of America
Dover Publications, Inc., 31 East 2nd Street, Mineola, N.Y. 11501

INTRODUCTION TO
THE DOVER EDITION.

BY JOSHUA STOFF, CURATOR
CRADLE OF AVIATION MUSEUM.

THERE is probably no single book that has contributed more to the history of aviation than Octave Chanute's *Progress in Flying Machines*. Certainly no book was more seminal in the birthing stage of man's flight, for it represented the first step toward a free exchange of ideas among aeronautical experimenters. *Progress in Flying Machines* provided aviation enthusiasts with a solid survey of almost every important experiment and invention relating to the problem of mechanical flight. His own efforts made Chanute the leader in a new era of aeronautical development, and his outspoken support and enthusiasm led many experimenters to appreciate the possibility of flight by man.

Chanute was born in Paris on February 18, 1832, although he was to spend most of his life in the United States. He came to America in 1838, when his father, a French professor of history, accepted the vice-presidency of a college in Louisiana; six years later the family moved to New York City. By 1849 Chanute was working as a railroad surveyor, his first project being the construction of the line between New York and Albany along the Hudson River. When the line reached Albany in 1853, he moved west to take the job of surveying and constructing a new line in Illinois. In the following years Chanute played a large part in the construction of numerous railroads in the American Mid- and Far West, earning distinction for his specialized work in bridge construction. He retired as an active engineer in 1888.

The year after his retirement Chanute moved from Kansas City to Chicago, where he turned his attention to aeronautics. Since the 1850s he had been intrigued by the possibility of man's flight, and he had collected data on the subject since he was about twenty-five. His readings in the centuries-long history of aeronautic experimentation gave him the idea that a careful study of past failures might aid in the eventual development of a practical flying machine. Thus moved to investigate the basic principles of flight, Chanute embarked on the very pragmatic approach of compiling data from other aeronautic experiments, so as to find out where he might begin his own research and in what direction it should best be pursued. By 1889 this work had become a full-time occupation for him.

Chanute's study of aeronautical history culminated in a series of articles published between 1891 and 1893 in the *Railroad and Engineering Journal.* These articles were reprinted in book form in 1894 as *Progress in Flying Machines.* This book was without question the most important work of its kind then in existence, for it consisted of a very concise study of nearly every experiment in aviation that had been made up to that point.

Progress in Flying Machines exerted a profound influence on everyone interested in the problem of mechanical flight. It gave reasonably accurate descriptions of successful inventions, and Chanute's explanations for the failures gave his colleagues a comprehensive review of past attempts to develop a flying machine. His analyses of the faults of earlier designs enabled experimenters to redirect their efforts from paths that had already proved dead ends, allowing them to concentrate instead on those principles that showed the most promise. Chanute's study helped prevent wasteful duplication, opening the way to more productive research.

As a result of his articles, Chanute was asked to organize the Third International Conference on Aerial Navigation. T' ⹀ conference was held in Chicago in 1893, and was attended by one hundred delegates from all over the world. It turned out to be a high point of the aeronautical awakening of the 1890s. Like Chanute's writings, it was noteworthy for fostering a spirit of cooperation and encouraging a free exchange of ideas among the world's leading aeronautical experimenters.

Chanute had come to realize by the mid-1890s that a sound scientific approach was needed to address the problem of mechanical flight. He believed that earlier experimenters had erred in focusing on building a lightweight engine, and that other, more important problems—such as controlling the machine and maintaining stability—had to be solved first before a practical flying machine could be developed. A lightweight engine would have made no difference between success and failure at this point.

No longer content just to study others' experiments, Chanute became eager to further the progress of aviation through his own practical work, and he decided to devote as much time as he could to this effort. Like the pioneer aviator Otto Lilienthal, he thought it important to first learn how to glide, and he had always recommended that his colleagues pursue this aspect first. Chanute decided to finance a series of gliding experiments himself, although he felt that his sixty-four years prohibited him from conducting them personally. He therefore secured the services of Augustus Herring, a young engineer and enthusiast, in 1895, and in 1896 they began to build two new types of glider.

That year saw Chanute, aided by Herring and two assistants, set up camp on the bluffs overlooking the southern shore of Lake Michigan, where he intended to test the machines he had designed. The multiwinged glider, with wings hinged to absorb the force of wind gusts, flew fairly well, but his triplane design worked even better. In this machine Chanute realized the idea of fastening the wings together with struts and wires—a method used in bridge construction—thereby giving the wings great strength and rigidity. This was the first successful application of the Pratt truss to the flying machine. When the machine was modified into a biplane it flew even better, exhibiting exceptional steadiness and control while in the air. Hundreds of flights were made with this glider in the summer of 1896.

The most important discovery that Chanute made with his gliders was how to maintain stability by moving the wings rather than the pilot of the aircraft. Lilienthal had maintained his aircraft's balance in the air by shifting the weight of his body within the frame of the machine; Chanute made the craft's surfaces movable instead of the man, and much of this movement was designed to take place automatically. He was clearly of the opinion that when the glider sailed through the air, its pilot should be comfortable and relaxed, devoting his attention to flying the aircraft rather than to the gymnastics many designs required simply to remain aloft. The credit for having first attempted to develop a naturally stable aircraft goes to Chanute.

These experiments represented the end of Chanute's active participation in aeronautics, although his numerous publications and speeches on the subject kept him occupied with the problem of mechanical flight. In retrospect he realized the enormous advancements that the previous decade had brought to aeronautics, and by 1898 he was able to write in the *Aeronautical Journal:* "Probably few readers realize the advance which has been made towards a practical solution of the problem of aviation in the past eight years. It is greater than all that was accomplished during the preceding two centuries. Not only has the subject been rescued from the contempt brought upon it by past failures and eccentric proposals, but very able men—Langley, Maxim, Pilcher, Hargrave and Lilienthal—have shown by their experiments and writings that man may fairly hope to fly through the air."

In March 1900 Chanute received a letter from Wilbur Wright, a young bicycle mechanic from Dayton, Ohio, who had studied *Progress in Flying Machines* with great care and enthusiasm. Wright sought the older man's advice concerning the materials and methods used in constructing gliders, as well as his insight on the selection of a good testing site. He told Chanute that for several years he had been "afflicted" with the belief that flight by man was possible, and that he and his brother Orville intended to devote some time to experimenting with gliders. Chanute, who never considered his discoveries to be secret, answered promptly, advising Wright on suitable locations for testing gliders, and citing earlier accounts of such experiments. These letters marked the beginning of a long and profitable friendship that would result in the construction of a practical airplane.

Soon a regular exchange of letters began between Chanute and the Wrights. This correspondence continued for more than a year before they first met in June 1901. By then the Wright brothers had completed and tested their first glider, the design of which was indebted to Chanute's biplane of 1896. Subsequent Wright designs were loosely based on Chanute's basic principles, and they modified his truss construction to suit their own purposes. Eventually Chanute became a regular visitor to the Wright camp at Kitty Hawk, North Carolina, always ready to share his knowledge.

Chanute spent a week at the camp in August 1901, and his observations of their glider convinced him that they had greatly improved upon his own design. The Wrights soon learned, however, that virtually all scientific data concerning lift was grossly inaccurate, and without dependable information to go on, they became discouraged and considered giving up experimentation. Chanute immediately talked them out of abandoning their tests, reassuring them that they had come closer to understanding the problem of mechanical flight than had anyone else. This encouragement may have been Chanute's greatest contribution to aeronautics, for the Wrights listened to his advice and continued their work.

On December 17, 1903, the Wrights finally flew the world's first practical airplane at Kitty Hawk. Chanute unfortunately was not there to witness the triumph, although he did observe later flights. He remained a devoted friend and advisor to the Wrights during the difficult years when several others were seeking to rob them of their invention, and he acted as an unofficial spokesman for the brothers until 1908. At a time when reports of the Wrights's successes were often met with disbelief and suspicion, Chanute confirmed and supported their results in his speeches and letters.

Chanute suffered a severe attack of pneumonia while in Paris in June 1910, and he died in November of that year. He had been fortunate enough to see in his lifetime the fruition of his dreams: namely, the success of aviation. For someone who had worked toward man's flight for much of his life, to actually see an airplane fly must have been a breathtaking, exhilarating experience. Its success was largely due to Octave Chanute himself, who changed the world of aviation—and our lives—forever.

GARDEN CITY, August, 1997.

INDEX TO ILLUSTRATIONS.

*Those marked with a * have been tested by experiment.*

PREFACE.

The following pages consist of a series of articles on "Progress in Flying Machines," as distinguished from balloons, which have been published in *The Railroad and Engineering Journal* (now redesignated as *The American Engineer*), of New York City.

The first article appeared in October, 1891, and the series comprised 27 issues. It was at first expected to explore the subject in six or eight articles, but investigation disclosed that far more experimenting of instructive value had been done than was at first supposed, and not only have these articles run to greater length than was expected, but they have been thought worthy of issuing in book form.

Naturally enough the public has taken little heed of the progress really made toward the evolution of a complicated problem, hitherto generally considered as impossible of solution. It will probably be surprised to learn how much has been accomplished toward overcoming the various difficulties involved, and how far the elements of a possible future success have accumulated within the last five years.

The writer's object in preparing these articles was threefold:

1. To satisfy himself whether, with our present mechanical knowledge and appliances, more particularly the light motors recently developed, men might reasonably hope eventua'ly to fly through the air. He now thinks that this question can be answered in the affirmative.

2. To save the waste of effort on the part of experimenters, involved in trying again devices which have already failed; and to point out, as much as may be, the causes of such failures. To this end an earnest effort was made to gather all the experimental records which were accessible, and to obtain a thorough understanding of them, so as to bring out clearly the reason of the failure. The reader must be the judge as to the measure of success which has attended this effort.

3. To furnish an account of those recent achievements which render it less chimerical than it was a few years ago to experiment with a flying machine, and to give such an understanding of the principles involved and of the results thus far accomplished, as to enable an investigator to distinguish between an inadequate proposal, sure to fail, and a reasonable design, worthy of consideration, and perhaps (after due investigation and preliminary trial) of experiment upon an adequate scale.

Chicago, January, 1894.

FLYING MACHINES.

INTRODUCTION.

HAVING in a previous volume treated the general subject of "Aerial Navigation," in which a sketch was given of what has been accomplished with balloons, I propose in the following chapters to treat of Flying Machines proper—that is to say, of forms of apparatus heavier than the air which they displace ; deriving their support from and progressing through the air, like the birds, by purely dynamical means.

It is intended to give sketches of many machines, and to attempt to criticise them.

We know comparatively so little of the laws and principles which govern air resistances and reactions, and the subject will be so novel to most readers, that it would be difficult to follow the more rational plan of first laying down the general principles, to serve as a basis for discussing past attempts to effect artificial flight. The course will therefore be adopted of first stating a few general considerations and laws, and of postponing the statement of others until the discussion of some machines and past failures permits of showing at once the application of the principles.

The first inquiry in the mind of the reader will probably be as to whether we know just how birds fly and what power they consume. The answer must, unfortunately, be that we as yet know very little about it. Here is a phenomenon going on daily under our eyes, and it has not been reduced to the sway of mathematical law.

There has been controversy not only about the power required, but about the principle or method in which support is derived. The earlier idea, now abandoned, so far as large birds are concerned, was that when they flapped their wings downward they produced thereby a reacting air pressure wholly equal to their weight, and so obtained their support. This is known as the "orthogonal theory," and has been disproved by calculations of the velocity and resulting pressures of the wing beats of large birds, and by the more recent labors of Professor Marey. It seems likely that the smaller birds, who, as will be explained hereafter, are probably stronger in proportion to their weight than the larger birds, possess the power of delivering blows upon the air equal to a supporting reaction. Such may be the case in the hovering of the humming-bird and the rising vertically of the sparrow ; but the latter exertion is evidently severe, and cannot be long continued.

Mr. Drzeweicki has shown that a buzzard, beating his wings 2½ times a second, with an amplitude of 120°, could only obtain, according to accepted formulæ of air pressures, a sustaining orthogonal reaction of 0.40 pounds or about $\frac{1}{10}$ of his weight, while if his wings are considered as inclined planes, progressing horizontally at a speed of 45 miles per hour, a sustaining reaction is easily figured out.

It seems quite certain that large birds cannot practice orthogonal flight, and that they derive their support mainly if not wholly from the upward reaction or vertical component of the normal air pressure due to their speed. That they are living Aeroplanes, under whose inclined wings their velocity creates a pressure which is normal to the surface. This is confirmed by the great difficulty which they experience in getting under way. They run against the wind before springing into the air, or preferably drop down from a perch in order to gain that velocity without which they cannot obtain support from the air. Thus the surfaces of their wings act as aeroplanes as well as propellers, the latter action being produced by the direction of the stroke and the bending upward of the rear flexible portion of the feathers.

Bird flight may be considered as comprising three phases :

1. Starting, during which great exertion must be made, unless gravity can be utilized.

2. Sailing, or flight proper, during which the bird exerts his normal force, or makes use of that of the wind, as will be more particularly explained hereafter.

3. Stopping, in which great exertion may again be required, if the headway is to be rapidly stopped, or in which the retarding force of gravity may be brought to do the work by simply rising to a perch.

Artificial flying machines will certainly have to conform to these three phases of flight, by providing methods of starting and stopping in addition to the means for performing the act of flight proper.

Birds perform all their manœuvers by regulating the intensity of their action, and by changing the angles at which they attack the air. Hence the important thing for us to know is to ascertain what pressure exists under a wing or, to simplify the question, under a plane surface, when it meets the air at a certain velocity and with a certain angle of incidence.

This has been, until the recent publication of Professor Langley's most important labors, a subject of uncertainty, which uncertainty he has done much to remove. We had had glimpses of the law ; but notwithstanding very many experiments by physicists, its numerical values were a subject of doubt and controversy among the few who gave any attention to the subject. It was the missing link, which rendered nearly unavailable the little that was known in other directions.

By the law of fluid reactions all air pressures are " normal," or exerted perpendicularly to the surfaces against which they bear ; now the question was : What is the relation between the pressure of a current of air of known velocity against a thin plane surface placed at right angles thereto, and the normal pressure of that same current against the same plane, if the latter be inclined to the current at an angle of incidence less than 90 degrees ?

Newton impliedly gave a solution ; but experiments long ago proved it to be wrong, although it is still taught in the schools and given in formulas in engineering reference books. He assumed, plausibly enough, that the proportional normal pressure was in the ratio of the sine of the angle of incidence, and when experiment showed this to be erroneous, other formulas were proposed, the following being a few of those which have been wrangled over :

Calling a the angle, and P the pressure on the inclined surface, while P' is that upon the right-angled surface, the following were assumed to represent the relation :

$$P = P' \sin a \qquad\qquad P = P' \sin^2 a$$
$$P = P' \sin^3 a$$
$$P = P' \frac{2 \sin a}{1 + \sin^2 a} \qquad P = P' (\sin a)^{1.84 \cos a}$$
$$P = 2 P' \sin a$$

Indeed, the field seemed so open in this direction that only two years ago I ventured to propose a formula of my own, which I subsequently concluded to be erroneous ; but the question seems now to be set at rest for the present by the experiments of Professor Langley, who proposes no formula of his own, but who shows that his results approximate very closely to the formula of Duchemin :

APPROXIMATE PERCENTAGES OF NORMAL PRESSURE. DERIVED FROM CHART OF EXPERIMENTS AND THEORIES. CALCULATED BY BOSSUT'S OR DUCHEMIN'S FORMULA.

$$P = P' \frac{2 \sin a}{1 + \sin^2 a}$$

Deg. of Angle.	Results of S. P. Langley's Experiments.	Proportion Normal Pressure.	Lift.	Drift.
1		0.035	0.035	0.000611
1½		0.052	0.052	0.00136
2		0.070	0.070	0.00244
3		0.104	0.104	0.00543
4		0.139	0.139	0.0097
5	0.15	0.174	0.173	0.0152
6		0.207	0.206	0.0217
7		0.240	0.238	0.0293
8		0.273	0.270	0.0381
9		0.305	0.300	0.0477
10	0.30	0.337	0.332	0.0585
11		0.369	0.362	0.0702
12		0.398	0.390	0.0828
13		0.431	0.419	0.0971
14		0.457	0.443	0.1155
15	0.46	0.486	0.468	0.124
16		0.512	0.492	0.141
17		0.538	0.515	0.157
18		0.565	0.538	0.172
19		0.589	0.556	0.192
20	0.60	0.613	0.575	0.210
21		0.637	0.594	0.228
22		0.657	0.608	0.246
23		0.678	0.623	0.264
24		0.700	0.639	0.286
25	0.71	0.718	0.650	0.304
26		0.737	0.662	0.323
27		0.752	0.670	0.342
28		0.771	0.681	0.362
29		0.786	0.686	0.382
30	0.78	0.800	0.693	0.400

Deg. of Angle.	Results of S. P. Langley's Experiments.	Proportion Normal Pressure.	Lift.	Drift.
31		0.815	0.698	0.421
32		0.828	0.702	0.439
33		0.843	0.706	0.459
34		0.853	0.707	0 478
35	0.84	0.867	0.708	0.498
36		0.878	0.709	0.516
37		0.885	0.709	0.532
38		0.894	0.705	0.551
39		0.902	0.701	0.569
40	0.89	0.910	0.697	0.586
41		0.918	0.693	0.602
42		0.926	0.688	0.619
43		0.934	0.683	0.638
44		0.941	0.676	0.654
45	0.93	0.945	0.666	0.666

$$P = P' \frac{2 \sin a}{1 + \sin^2 a}$$

I had already independently reached a conclusion quite similar. Finding that my formula was incorrect, I had a chart plotted, on which were delineated all the experiments on inclined surfaces which I could learn about—those of Hutton, Vince, Thibault, Duchemin, De Louvrié, Skye, the British Aeronautical Society, and W. H. Dines ; and on this chart I also had plotted the curves of the various formulas. The whole exhibited great discrepancies, yet by patient analysis various probable sources of error were eliminated, and the conclusion was reached that the formula last given, which I have seen variously attributed to Bossut or to Duchemin, was probably correct.

From this formula I had computed for my own use the accompanying table of normal pressures ; and as it seems to be quite confirmed by Professor Langley's experiments, and seems to promise to be of great use, I now venture to publish it.

Once the normal pressure is known at a particular angle of incidence, its static components in different directions can be obtained by the laws governing the resolutions of forces. This was shown, as early as 1809, by Sir George Cayley, in the following demonstration, in which he ingeniously evades the then prevailing confusion about the "law of the angle" by starting with the weight of the bird instead of its wing surface and velocity.   He says :

When large birds, that have a considerable extent of wing compared with their weight, have acquired their full velocity, it may frequently be observed that they extend their wings, and, without waving them, continue to skim for some time in a horizontal path.

Fig. 1 represents a bird in this act.   Let *A B* be a section of the plane of both wings. opposing the horizontal current of air (created by its own motion), which may be represented by the line *C D*, and is the measure of the velocity of the bird.   The angle *B D C* can be increased at the will of the bird, and to preserve a perfectly horizontal path, without the wing being waved, must continually be increased in a complete ratio (useless at present to enter into), till the motion is stopped altogether ; but at one given time the position of the wings may be truly represented by the angle *B D C*.   Draw *D E* perpendicular to the plane of the wings. produce the line *C D* as far as required, and from the point *E*, assumed at pleasure in the line *D E*, let fall *E F* perpendicular to *D F* ; then *D E* will repre-

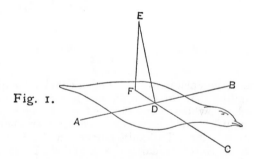

Fig. 1.

sent the whole force of the air under the wing—*i.e.*, normal pressure, which being resolved into the two forces *E F* and *F D*, the former represents the force that sustains the weight of the bird, and the latter the retarding force by which the velocity of the motion producing the current *C D* will be continually diminished ; *E F* is always a known quantity, being equal to the weight of the bird. and hence *F D* is also known, as it will bear the same proportion to the weight of the bird as the *sine* of the angle *B D C* bears to its *cosine*, the angles *D E F* and *B D C* being equal.

In the table, pages 4 and 5, the first column shows the degree of the angle of incidence : the second the result of Professor Langley's experiments ; the third the proportion or percentage which the normal pressure at that angle bears to the pressure at the same velocity of the same plane at right angles to the current ; while the fifth and sixth columns show the resolutions of this normal pressure, being the force which sustains the weight of the bird vertically as against gravity, which is here termed the " Lift ;'' and the retarding force against horizontal motion, which is here termed the " Drift.'' They are calculated by multiplying the normal pressure by the *sine* and by the *cosine* of the angle.

In order to obtain the aggregate normal pressure, or the lift and the drift, upon any thin plane surface, it is simply necessary to multiply its area by the pressures per square foot, which are given (approximately) in the ordinary tables of wind velocities, and this again by the percentages given in the table.

The angles are only given up to 45°, as more than this would be useless to the general reader; and it will be noted that there is an angle of maximum uplift at about 36°. This results from the fact that the normal pressure is continually increasing, while the *cosine* of the angle is continually diminishing, but not equally, so that their product reaches a maximum, as stated. This is confirmed by the results of Professor Langley's experiments, as recorded on page 58 of his " Experiments in Aerodynamics.''

It should be borne in mind that the table only purports to apply to thin planes one foot square, and hence is given as containing only approximate percentages of normal pressures. For other shaped planes, for curved surfaces, and for solids the percentages may be different, because a great many anomalies have been found in experimenting upon air resistances, and we yet know painfully little about them.

For instance, the following may be mentioned :

1. For high velocities, such as those of projectiles, the resistances do not vary as the square of the speed, as assumed in ordinary tables ; they more nearly approach the cube of the velocity.

2. If a thin plane be exposed to a current of air, at right angles thereto, the pressure on the plane is not uniform over all its surface, but is greatest at the center.

3. Plane surfaces of equal areas but of different shapes (square, oblong, triangular, etc.) are found to receive slightly different pressures at the same speed. Moreover, the average pressure per square foot varies with mere variation of size on the same shaped planes.

4. The pressure upon an inclined elongated surface will vary for the same speed, whether it be exposed longitudinally or transversely to the current.

5. Holes may be cut in thin planes without reducing the aggregate pressure in proportion to the surface cut away.

Moreover, the aggregate pressure may be made to vary by simply changing the position of the holes.

6. Inclined planes may be superposed without diminishing the sum of their separate individual pressures, provided they are properly spaced with regard to the angle of incidence. If too close, they will interfere with each other, but the amount of such interference will vary with the speed.

7. Perfectly horizontal planes, free to fall, have their time of falling much retarded if in rapid horizontal translation.

8. The weight remaining the same, the force requisite to sustain *inclined* planes in horizontal motion diminishes instead of increasing, when the velocity is augmented.

9. If the plane be gradually inclined to the current, the point of maximum pressure will move forward toward the front edge as the angle of incidence diminishes. The position as given by Joëssel's law is shown by the formula :

$$C = (0.2 + 0.3 \sin a) L,$$

in which C represents the position of the center of pressure, L the length, and a the angle of incidence, the formula indicating that the position of the center of pressure varies from 0.5 to 0.2 of the distance from the front to the center of the plane.

Of these anomalies, the 6th, 7th and 8th were experimentally determined by Professor Langley ; and he partly confirmed the 9th, as well as giving strong confirmation to the results of Duchemin on the " law of the angle" previously mentioned. The 8th is especially important, and its consequences are pointed out by Mr. Langley in the following words :

The most important general inference from these experiments, as a whole, is that, so far as the mere power to sustain heavy bodies in the air by mechanical flight goes, such mechanical flight is possible with engines we now possess, since effective steam-engines have lately been built weighing less than 10 lbs. to 1 H.P., and the experiments show that if we multiply the small planes which have been actually used, or assume a larger plane to have approximately the properties of similar small ones, 1 H.P., rightly applied, can sustain over 200 lbs. in the air, at a horizontal velocity of over 20 meters per second (about 45 miles an hour), and still more at still higher velocities.

These general remarks chiefly apply to thin plane surfaces, such as might be used in flying machines, but mere thickness plays an important part ; for in a solid body, with the same area of exposed head surface, the pressure will be varied by the depth, and especially by the form of the body in the rear. Thus curved surfaces and solids have quite different coefficients of pressure from thin flat planes, and theoretical estimates of their resistances have hitherto proved to be quite wrong.

Indeed, it may be said with respect to curved surfaces and solids, that a glimpse has been caught of a still more mysterious phenomenon. It is known that certain shapes, when exposed to currents of air under certain ill-understood circumstances, actually move toward that current instead of away from it. Thus a hollow sphere impinged upon by an air jet will move up toward it instead of away. The lower disk in Professor Willis's apparatus, when blown upon, moves against the current toward the upper disk. Dr. Thomas Young proved, in 1800, that a certain curved surface suspended by a thread approached an impinging air current, instead of receding from it. M. Goupil found, in experimenting, that a suspended hollow shape was first blown out to a horizontal position by a wind of sufficient velocity, and then, when that velocity increased, actually drew into the wind for an instant and slackened the tension on the cord. It is also said that certain forms of windmills wear more on the front stop than on the back stop of their axle of rotation ; so that there seems to be a mysterious action, which some French observers, who have been watching birds soar, have, for want of a better term, called their " Aspiration," by which a body acted upon by a current may actually draw forward into that current against its direction of motion.

Thus it is seen that in such complicated matters theory cannot progress in advance of experiment, and the extreme importance of those experiments hitherto tried, or hereafter to be tried by a physicist possessing the ability of Professor Langley, will in part be appreciated.

Science has been awaiting the great physicist, who, like Galileo or Newton, should bring order out of chaos in aerodynamics, and reduce its many anomalies to the rule of harmonious law. It is not impossible that when that law is formulated all the discrepancies and apparent anomalies which now appear, will be found easily explained and accounted for by one simple general cause, which has been hitherto overlooked.

Thus far, Professor Langley seems to have experimented upon plane surfaces only, and to have measured chiefly what has been termed in the table here given the " lift" and the " drift" at various angles. His conclusions therefrom are very important ; but the " drift" will not be the sole resistance to be encountered, for the sustaining surfaces of a flying machine must not only have a certain thickness, to give them the necessary strength and rigid-

ity, but there will be friction of air upon them, and there
must be a solid body or hull to contain the machinery
and the cargo.

Thus the elements of resistance are three in number :
1. The hull resistance.
2. The drift.
3. The skin friction.

Of the skin friction Professor Langley says that it is apparently so small that it may be neglected without material error ; and he has given the measure of the " drift"
as the result of his experiments.

The head or hull resistance will probably be found to be
the chief element which will limit the possible speed of
flying machines. It will probably grow as the square of
the velocity, thus requiring the power exerted to vary as
the cube of the speed, but will be modified by a series of
coefficients, due to the shape of the solid body, just as
some birds are swifter flyers than others of the same
weight, in consequence of their difference in shape.

Hence the power required to drive such a machine can
only be approximated at present ; but this will be more
particularly discussed when treating of the areas of supporting surfaces and speed of birds, for the reader may be
impatient to be told something of what has been attempted
by man.

Inventors, in their ignorance of the laws of air reactions
and resistances, have proposed all sorts of devices for
compassing artificial flight and experimented with not a
few ; so that Mr. E. Dieuaide, of Paris, upon making a
study of the subject, published in 1880 an illustrated
chart,* in which he delineated the more remarkable machines which had been proposed for aerial navigation without the use of balloons. This chart contains some 53
figures ; and from this, as well as from the book of M.
Gaston Tissandier on Aerial Navigation,† which contains
much accurate information, the following has been chiefly
compiled, in which it will be attempted not only to give an
account of what has been proposed, so far as the meager
data will permit, but also to critcise the machines with the
light of our present knowledge, and to endeavor to point
out why they failed. Failures, it is said, are more instructive than successes ; and thus far in flying machines
there have been nothing but failures.

These various machines, diverse as they are, may rough-

---

* *Tableau d'Aviation.* Représentant tout ce qui a été fait de remarquable
sur la navigation Aérienne sans Ballons. Published by the author.
† *La Navigation Aérienne.* Par Gaston Tissandier. Hachette et Cie ;
octavo, 334 pp.

ly be classed under the three following heads, according to the intentions and theories which were held by their authors, as to the most efficacious way of deriving support from the air.

A. Wings and parachutes.
B. Screws to lift and propel.
C. Aeroplanes.

A.—WINGS AND PARACHUTES.

THE earlier adventurers upon aerial enterprises possessed little accurate knowledge of the properties of air. They had only their observations of the birds as a guide, and knew of no motive power save that derived from muscular energy ; hence their thoughts first turned to flapping wings, to be propelled by their own exertions. Some few, as we shall see, have considered the force of the wind, but it is only since the age of steam that artificial motors of any kind have been proposed for flying machines.

The well-worn legends of antiquity, concerning *Dedalus, Abaris, Archytas,* etc., may be passed over without comment. They merely indicate how the problem of artificial flight appealed to the imagination of men from the earliest periods, but some curious traditions will be mentioned, indicating partial successes in soaring flight, when we come to treat of aeroplanes.

About the first authentic account which we have of a proposal to provide man with flapping wings seems to be due to *Leonardo da Vinci,* the painter, sculptor, architect, and engineer. He is said not only to have experimented with aerial screws made of paper, and to have designed a

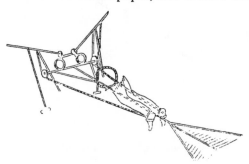

FIG. 2.—LEONARDO DA VINCI—1500.

parachute, but also to have seriously contemplated building an apparatus to propel a pair of wings, of which several sketches have been found in his note-books.

The first sketch shows a wing, actuated by the arms,

but *Da Vinci*, becoming aware, upon reflection, that all possible muscles of man must be brought into play to act effectually upon the air, designs in the second and third sketches an apparatus in which the wings are to be waved downward by the legs and lifted up by the arms. The third sketch is represented in fig. 2. In this *Da Vinci* only shows the legs in place, so as not to obscure the construction of the parts. The date is probably about the year 1500. The construction is simple, and might not prove altogether inefficient did the muscles of man possess the same energy and rapidity of action as do those of birds in proportion to their respective weights. It is not known just how far *Da Vinci* elaborated his idea, but he never put it to practical test, and it is chiefly mentioned here as a curious forerunner of actual experiments.

The first wing experiment reported by tradition seems to be that of a French tight-rope dancer named *Allard*, who, under the reign of Louis XIV., announced that he would fly from the terrace at Saint Germain toward the woods of Vesinet in presence of the king. It is probable that he had previously succeeded in gliding short distances, but upon trial before the court his strength failed him ; he fell near the foot of the terrace, and he was grievously hurt.

This probably occurred about the year 1660, and in 1678 a French locksmith named *Besnier* constructed a

FIG. 3.—BESNIER—1678.

pair of oscillating wings, approximately represented in fig. 3.

The apparatus consisted of two bars of wood hinged over the shoulders, and carrying wings of muslin, arranged like folding shutters, so as to open flat on the down stroke and fold up edgewise on the up stroke. They were alternately pulled down by the feet and by the arms, in such wise, that when the right hand pulled down the right wing, the left leg pulled down the left wing, and so on, thus imitating the ordinary movements in walking.

Besnier did not pretend that he could rise from the ground or fly horizontally through the air. He only tried short distances ; having begun by jumping off from a chair, then from a table, then from a window-sill, and next from a second story, and finally from the garret, on which occasion he sailed over the roof of an adjoining cottage. He gradually grew more expert, sold his first pair of wings to a mountebank, who performed with them at the fairs, and he expected with his second pair to fly across moderately wide rivers by starting from a height, but it is not known whether he ever performed this feat.

The illustration is evidently an imperfect sketch made from a description ; for the hinging at the shoulder is not shown, the attachment for pulling down the wings with the legs is evidently inefficient, and the supporting surfaces are entirely inadequate. The four wings are apparently each 3 ft. by 2 ft ; say, an aggregate of 24 sq. ft. in area, while in the table of birds, to be given hereafter, it will be seen that the duck, which has the smallest bearing surface in proportion to its weight, measures 0.44 sq. ft. to the pound, and at this rate a man, weighing, say, 150 lbs., would require wings aggregating 66 sq. ft. in area. It is probable that *Besnier* had even more than this, that he took short downward flights aided by gravity, but that he utterly failed when he undertook to go considerable distances.

It is not stated whether the *Marquis de Bacqueville* had engaged in similar preliminary practice when he announced, in 1742, that he would, on a certain day, fly across the river Seine from his mansion, situated in Paris on the quay at the corner of the Rue des Saints Pères, and alight in the Tuilleries, a distance of 500 or 600 ft. A large crowd having assembled on the appointed day, the marquis, with large wings attached to his hands and to his feet, launched himself into space from the summit of a terrace jutting out from one side of the mansion.

For a space he seemed to get along well, but soon his movements became uncertain, he faltered, and then he fell, alighting upon the deck of a washerwomen's barge a short distance out into the stream. He broke his leg in the fall, and never attempted the feat again.

The *Marquis de Bacqueville* was judicious in trying the experiment over a water-bed, for could he have held out but a few feet further he would probably have escaped with a mere ducking. He probably glided about 120 ft. with most violent exertions, and fell when his strength became exhausted. Fig. 4, which is probably incorrect,

represents the traditional apparatus with which this feat
was attempted. The surfaces measure about 24 ft. in
area, and are quite insufficient to sustain the weight of
a man.

Aware of this experiment of *De Bacqueville* and of its
consequences, the *Abbé Desforges*, a canon of the church at
Sainte-Croix at Étampes, invented, in 1772, a flying chariot,

FIG. 4.—DE BACQUEVILLE—1742.

with two wings and a small horizontal sail or aeroplane attached, which from contemporary descriptions seem to
have measured about 145 sq. ft. in aggregate area. He
expected to rise from a height of a few feet above the
ground, and to fly horizontally by rapidly beating his
wings. Upon actual trial, the machine being held aloft
by four men, the *Abbé* flapped violently, but utterly failed
to start off. Indeed, some of the accounts say that the
action of the wings pulled him down instead of up, so
that he got a harmless tumble when the men let go.

In 1781, *Blanchard*, who subsequently became a fervent
aeronaut, and who was the first to cross the British Chan-
nel in a balloon, constructed near Paris a flying chariot
with four wings, measuring in the aggregate some 200
sq. ft. in area. He never exhibited the apparatus in public, having probably ascertained by private experiment
that he was unable to move the wings rapidly enough to
produce any useful effect.

These last two experiments, taken in connection with
those previously mentioned, exhibit fairly well the two
horns of the dilemma that confront inventors who endeavor to provide man with wings to be worked by his
own muscular power. Either those wings have to be
relatively small, in order to permit their being waved rapidly—and then they do not afford sufficient supporting
area—or if they are made to approximate to the proportion
which generally obtains with birds, or about one square
foot to the pound, they become so large that the man does
not possess the muscular power to wave them at any
adequate speed.

Ideas, however, die hard, and we may disregard somewhat the chronological order of date, in order to follow the evolution of the small-wing idea, which each fresh inventor fancies has been incorrectly worked out by his predecessors.

Of these was *Bourcart*, who in 1866 experimented with the apparatus shown in fig. 5 It consisted of four wings with a feathering action, so that it presented the edge to the air upon the up stroke and the broad side upon the

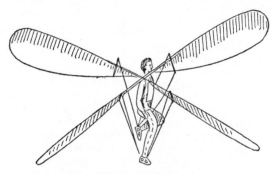

FIG. 5.—BOURCART—1866.

down stroke, but the results were insignificant, and the experiment was abandoned. The supporting areas measure approximately some 36 sq. ft., but are only effective upon the down stroke.

In 1873 Professor *Pettigrew* published his work on " Animal Locomotion," in which he called attention to the fact that birds in flapping flight, flex their wings so as to resemble a screw propeller, and that the tips describe a figure of 8 motion. This led to the inference that man had not succeeded in raising himself with wings because

FIG. 6.—DANDRIEUX—1879.

he had not hit upon the right motion, and in 1879 *Dandrieux* constructed an apparatus in which the wings were attached to an oblique axle, so as to describe a figure of 8 movement. This is represented in fig. 6, and there being but two wings in place of four, the supporting surfaces measure about 32 sq. ft. in area. The result was not satisfactory ; a partial alleviation of the weight was obtained, but nothing like human flight or the hope of it.

A much more successful experiment had, however, previously been made at the first Exhibition of the Aeronautical Society of Great Britain, held at the Crystal Palace, in London, in 1868. Mr. *Charles Spencer* exhibited an apparatus consisting of a pair of wings measuring each 15 sq. ft. in area, to which was attached an aeroplane measuring 110 ft. more, and also a tail like a boy's dart, and a longitudinal keel-cloth to preserve the equilibrium, the whole weighing 24 lbs. and giving a sustaining surface of 140 sq. ft. As Mr. *Spencer* was an athlete, he was enabled, by taking a preliminary run down a little hill, to accomplish short horizontal flights of 120 to 130 ft., in which he was wholly sustained by the air. He weighed 140 lbs., and his apparatus, which, it will be noted from the description, differed from those which propose " wings for man" by the addition of an aeroplane, measured 0.85 sq. ft. to the pound, or about the proportion of the larger soaring birds. The experiments attracted great attention at the time, but were not sufficiently encouraging to warrant pursuing the matter further.

At the same exhibition Mr. *W. Gibson* showed a machine consisting of two pairs of wings, worked by the hands and feet together, so as to impart a feathering movement similar to that of birds. He stated that in a former machine, having only one pair of wings of lighter construction, their action upon the air during a vigorous down stroke was sufficient to raise the man and machine ; but no practical demonstration was given, and although the inventor stated that he was then engaged in constructing a more perfect machine, nothing more has been heard of it.

Notwithstanding these many failures, the idea does not seem to be dead yet, for in September, 1890, Mr. *W. Quartermain*, who exhibited an explosion engine for aerial purposes in 1868, in which the motive power was derived from the gases generated from a species of rocket composition, wrote a letter to the London *Engineer*, in which he stated that he had abandoned his attempts to procure a light and energetic motor from hydrocarbonous matter, in favor of man's weight and muscular power, which he considers preferable, and was then engaged in experimenting with an apparatus consisting of four wings, formed after the stag beetle type, each 10½ ft. long by 2¼ ft. wide, opposing 90 sq. ft. of expanse of surface to the air. This arrangement weighed 350 lbs., including 212 lbs. for the weight of the operator, who by working both handles and treadles, thus bringing all his muscles into action as well as his weight, was enabled to wave the wings, which are 25 ft. from tip to tip, so as to produce a double stroke for every single stroke of his body on the motive shaft. He describes the result as resembling that of domestic fowls flapping their wings without lifting themselves from the ground, but is of opinion that the uplifting force was greater than his weight of 212 lbs., and believes that further improvements in the mechanism, with more skilful workmanship, might produce an ascensive force greater than the whole weight of 350 lbs. This may well be doubted, for not only will it be shown hereafter that the energy of man must be less than that of birds, but none of the latter fly with so small a bearing surface in proportion to the weight—0.26 square foot to the pound—as in *Quartermain's* apparatus.

It has been suggested, however, that umbrella-like surfaces might prove more effective than wings, and increase the uplift to be derived from the air. Such contrivances were experimented upon by Sir *George Cayley*, who constructed, about 1808, a pair of wings which appear from the drawings to have been a fabric stretched tightly over a dished frame, this framework consisting of two ribs at right angles to each other, bent and tied across so as to secure rigidity. This double umbrella contained 54 sq. ft. and weighed only 11 lbs., and the inventor says : " Although both these wings together did not compose more than half the surface necessary for the support of a man in the air, yet during their waft they lifted the weight of 9 stone" (126 lbs.). It is not stated with what speed they were wafted nor with what power, but that the result did not promise to provide " wings for man" may be inferred from the fact that Sir *George Cayley*, in a very valuable series of articles in *Nicholson's Journal* for 1809 and 1810, starts out with the assertion that, in order to accomplish aerial navigation, " it is only necessary to have a first mover which will generate more power in a given time, in proportion to its weight, than the animal system of muscles."

The next experiments with umbrella-like wings attracted attention all over Europe. They were carried on by *J. Degen*, a clockmaker of Vienna, from 1809 to 1812, with the apparatus shown in fig. 7. It consisted of two wings 8½ ft. wide and 22 ft. across in the aggregate, each being shaped somewhat like a poplar or an aspen leaf. They were stretched upon an umbrella-like frame and thoroughly braced back, both above and below, to a central stick by a number of small cords. The supporting

surfaces consisted of bands of taffeta so attached as to
have a valvular action, in order to imitate the supposed
action of the feathers of birds, and the total supporting

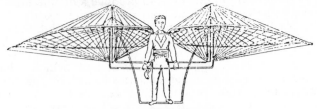

FIG. 7.—DEGEN—1812.

surface was 130 sq. ft., while the weight, without the
operator, was stated at 20 lbs.

With this apparatus *Degen* was stated, in 1809, to have
risen to a height of 54 ft., by beating his wings rapidly, in
presence of a numerous assembly in Vienna, and all the
newspapers began to publish accounts of the performance.

These descriptions failed to mention one important addition. *Degen* was also attached to a small balloon capable of raising 90 lbs., so that the uplift exerted by the
wings was only 70 lbs. of the 160 lbs. weight of the operator and his apparatus.

In 1812 *Degen* went to Paris to exhibit his invention.
He then stated that the balloon was of no sort of utility in
obtaining headway, but that it was necessary as a counter-
poise to maintain his equilibrium and to lighten his
muscular efforts. He evidently expected by the action of
his wings to drag the balloon along in still air while it
lifted part of his weight.

He gave three public exhibitions in Paris, but unfortunately for him, as there was wind upon each occasion, he
was blown away, and on the third attempt he was attacked
by the disappointed spectators, beaten unmercifully, and
laughed at afterward as an impostor.

The umbrella idea had, however, previously proved to
be of value for parachutes, and in 1852 *Letur* devised the
apparatus shown in fig. 8, with which he expected to
direct himself through the air, by means of the wings and
tail, first starting from an elevation.

In 1854 he ascended from Cremorne Gardens in London,
suspended about 80 ft. below a balloon manœuvred by
Mr. Adam, the areonaut, who was assisted by a friend.
Letur performed several evolutions in the air by means of
his wings, none of them apparently very conclusive; but
in coming down near Tottenham, the wind carried the
apparatus violently against some trees, and poor *Letur*
received injuries which resulted in his death.

His apparatus measured about 660 sq. ft. in bearing
surface, and had he been entirely detached from the balloon, it is possible that he might have reached the ground
in safety ; but it is evident that his wings would have been

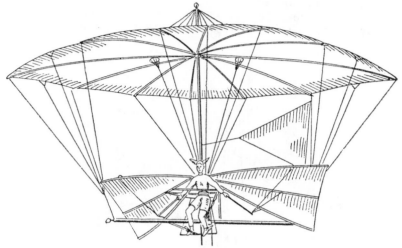

FIG. 8.—LETUR—1852.

of little service in enabling him to obtain more than a
slight horizontal direction.

Undeterred by this sad fate, a Belgian shoemaker
named *De Groof* designed, in 1864, an apparatus which
was a sort of cross between beating wings and a parachute.
His plan was to cut loose with it from a balloon, and to
glide down in a predetermined direction by manœuvring
the supporting surfaces. He endeavored to make a
practical experiment, both in Paris and in Brussels,
but it was only in 1874 that he succeeded in doing so in
London.

The apparatus is shown in fig. 9. It consisted of two
wings, each 24 ft. long, moved by the arms and the
weight of the operator, and of a tail 20 ft. long, which
could be adjusted by the feet.

De Groof first went up on June 29, 1874, from Cremorne
Gardens, London, attached to the balloon of Mr. Simmons. He came down safely, and claimed to have cut
loose at a height of 1,000 ft., but it was subsequently
stated by others that in point of fact he had not, upon
this occasion, cut loose at all, but had descended still
attached to the balloon. In any event, he went up again
on July 5 following, with the same aeronaut, and on this
occasion he really did cut loose.

The result was disastrous. As soon as, in the descent, pressure gathered under the moving wings, they were seen to collapse together overhead and to assume a vertical position, when *De Groof* came down like a stone, and was killed on the spot.

Had the wings been prevented from folding quite back, by means of suitable stops, the descent might not have

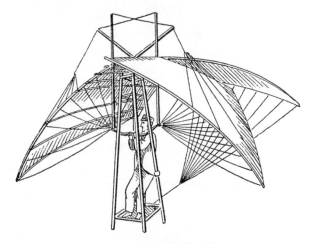

FIG. 9.—DE GROOF—1864.

proved fatal. The area of the wings and tail, as extended horizontally, is said to have amounted to 220 sq. ft., while the weight of the man and machine was 350 lbs., or at the rate of 0.65 square foot to the pound. This corresponds to a pressure of 1.54 lbs. to the square foot, which would be generated by a velocity of 25.7 ft. per second, or a free fall from a height of 10.3 ft.; an unsafe distance for an ordinary person, but not for a trained acrobat.

Ordinary parachute practice is said to allow from 2 to 3 sq. ft. per pound, corresponding to velocities in falling of 14.7 to 12 ft. per second.

It was the most egregious folly for *Letur* and *De Groof*, as well as for *Cocking*, who was killed in 1836 in an experiment with a parachute shaped like an inverted umbrella, to attempt a descent with an apparatus previously untried to test its strength and behavior. A few prior experiments, with a bag of sand, instead of the man, would have exhibited the action that was to be expected.

Another class of inventors of "wings for man" have endeavored to secure safety by the use of large bearing surfaces. The first of these was probably, *Meerwein*, architect to the Prince of Wales, in 1784, who proposed an apparatus shaped like the longitudinal section of a spindle, separated into two wings, by a hinge at the center. It measured nearly 200 sq. ft. in area, and probably was never tried, but if it had been, it is quite certain that a man could never have imparted to the wings sufficient velocity to perform any useful effect.

The next proposal of this class was that of *Bréant*, who

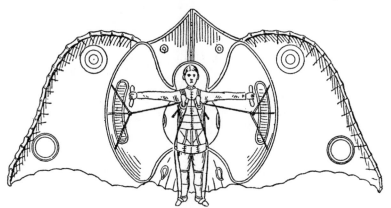

FIG. 10.—BRÉANT—1854.

designed in 1854 the apparatus shown in fig. 10. It consisted of two wings, each measuring about 54 sq. ft. in area, and provided with three valves to relieve pressure on the up stroke. The down stroke was to be produced by the joint action of the feet and hands, and the wings were to be drawn back by elastic cords. It is not known whether it was ever tried, but it would have proved ineffective if it had been.

The next design was that of *Le Bris* in 1857, which is exhibited by fig. 11. By noting the little man working the levers in the center, the proportions of the apparatus, which seems to have measured some 550 sq. ft. in area, will be appreciated. It is said to have been experimented with in a small model, in which levers pulled down the wings which were then drawn back by springs, but it did not succeed in rising into the air, as was hoped by the inventor.

Before proceeding to describe other designs for winged machines, to be driven by artificial motors instead of muscular power, it may be well to call attention to the fact that not only has every attempt of man to raise himself on the air by his own muscular efforts proved a complete failure, but that there seems to be no hope that any

amount of ingenuity or skill can enable him to accomplish
this feat.

It has been argued that there is no proof that, weight
for weight, a man is comparatively weaker than a bird,
and that, inasmuch as he can raise his weight in walking
up a stairway, he should be able to raise it by acting upon
the air with a suitable apparatus.   The weak point about

FIG. 11.—LE BRIS—1857.

this argument is not only that the weight and bulk of such
an apparatus become a surcharge on the muscular power
of the man, as would be, for instance, the case were an
artificial pair of wings applied to an ostrich, but that
among the birds themselves the power to rise vertically
unaided does not exist for the larger species.   These have
to resort to various artifices, such as running against the
wind or dropping from a perch, in order to gain that
initial velocity which enables their surfaces to derive support from the air, and this probably furnishes a good
reason why no flying birds exceed some 50 lbs. in weight ;
for small animals must possess more energy in proportion
to their size than large ones.

Assuming that the speed of contraction in the muscles
of two similar birds of different sizes is the same, it is
evident that the work done per unit of time will be in ratio
to the sectional area, or as the square of the dimensions,
while the weight to be moved will vary as the cube of the
dimensions ; hence the rate of increase between the energy
and the weight will be :

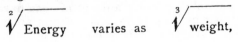

$$\sqrt[2]{\text{Energy}} \quad \text{varies as} \quad \sqrt[3]{\text{weight,}}$$

or to put it in the shape of formulas which shall express
the relative energy ot animals of the same class :

$$E \propto \sqrt[3]{w} \quad \propto \sqrt[1.5]{w} \quad \propto W^{\frac{2}{3}}$$

These being all merely different ways of writing it.
Hence we see that the energy of birds will only increase
as the $\frac{2}{3}$ power of their weight, and that there will be
an increase of size beyond which they will not be able to
develop the work required for a start.*

But man is also at a further disadvantage. Not only
do birds have an enormous muscular development, but
their muscles contract at a much more rapid rate than
those of other animals. Were men, therefore, not already
relatively weaker than smaller animals, in consequence
of the physical law which has been stated, they would still
be unable to develop energy fast enough to rise on the air
with a pair of wings. They can raise their weight, it is
true, but not as quickly as the birds. They can run up a
stairway at the rate of about 3 ft. per second, while the
sparrows rise up vertically at thrice that speed, and fly horizontally at 22 ft. per second.

While the inventors who experimented with flapping
wings, with which they tried to raise themselves on the
air by muscular effort, doubtless had it in mind eventually
to substitute artificial motors, if only they could catch the
trick by which the bird flies, there have been a few others
who have at the outset designed flapping wings, to be
moved by some primary artificial motor. As they generally knew of no such motor, within admissible limits of
weight in proportion to its energy, such designs have remained mere projects, and but few experiments have been
made.

The proposal of *Gérard*, in 1784, shown in fig. 12,
seems to have been among the first. It apparently provides, in addition to the body and wings, for a steering
arrangement in front, and for feet with springs to land
upon. The inventor omitted to state in his printed description what motive power he intended to use, but an
inspection of the drawing suggests the conjecture that the
apparatus was to be propelled in part by escaping gases,
like a rocket, and in part by flapping the wings through
the medium of a gunpowder engine ; proposals and experiments with such motors antedating, as is well known,

---

* Thus a bird of 50 lbs. weight can do no more work in a given time than
$50^{\frac{2}{3}} = 13.57$ similar birds each weighing 1 lb., or a bird of 1,000 lbs., did such a
one exist, could only develop the same number of foot-pounds per minute as
the aggregate of 100 analogous birds, each of 1 lb. weight.

those with the steam-engine. Be this as it may, soon after the success of the locomotive engine on the Liverpool & Manchester Railroad, Mr. *F. D. Artingstall* endeavored to compass an aerial locomotive. He con-

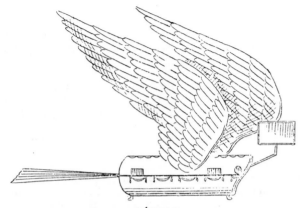

FIG. 12.—GÉRARD—1784.

structed a very light steam-engine, suspended it by a cord from the ceiling, and to the piston-rod he attached wings, which were so constructed that they opened somewhat like a Venetian blind on the up stroke and closed during the down stroke, moving through an arc of 80°. When steam was turned on the wings worked vigorously, but the machine jerked up and down, rushed from side to side, and, in fact, performed all kinds of gymnastic movements except flight. This experiment was terminated by the explosion of the boiler, and a second attempt, in which it was intended to use four wings instead of two, in order to keep up a continuous buoyancy, resulted in a second explosion ; after which the experiments were abandoned. In 1868 Mr. *Artingstall*, in a communication to the Aeronautical Society of Great Britain, stated the weak point in his various experiments to have been the lack of suitable equilibrium.

Every experimenter with aerial apparatus has doubtless encountered the difficulty of obtaining in a machine that equilibrium which the bird maintains by instinct, and also of deriving continuous support from the flapping of one pair of wings. These are probably the reasons which led *Struvé* and *Telescheff* to design, in 1864, the apparatus shown in fig. 13, in which five pairs of wings are attached to a central plane. The only description accessible to the writer states that the wings were moved by human force acting upon a spring, but it is evident that the apparatus was intended to be driven by artificial power, if the designers could only find one sufficiently light for that purpose. That they did not succeed in this

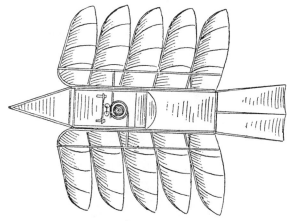

FIG. 13.—STRUVÉ & TELESCHEFF.—1864.

seems to be a fair inference from the fact that the machine was not tested by experiment.

At the Exhibition of 1868, of the Aeronautical Society of Great Britain, Mr. *I. Palmer* exhibited a pair of wings (to be driven by power) attached to a rotating axle, and so arranged that they expanded in the descent and closed in the ascent, like the action of a duck's foot in swimming ; this motion being obtained in a remarkably simple manner by a roller running on an eccentric cam, which could be instantaneously changed in position, so as to convert the vertical lifting power into one of horizontal force. It does not seem to have been applied to any flying machine.

At the same exhibition Mr. *I. M. Kaufmann*, engineer of Glasgow, exhibited the working model represented in fig. 14, which was intended as the precursor of an aerial steam machine weighing 7,000 or 8,000 lbs. The apparatus consisted of a steam boiler and engine, mounted upon wheels, and propelled by two long wings, which, during the down stroke, were set at an inclined direction backward, and were caused to turn at a forward angle during the up stroke. The main portion of the weight was to be sustained by superposed aeroplanes, and hence the machine should perhaps be described under that head, but it is here included under the head of wings, because of the mode of propulsion. The model weighed 42 lbs., and during the experiments with it its boiler, owing to its

small size, was not fired, steam being supplied from an
independent boiler. With steam pressure at 150 lbs. to
the inch, the wings made a short series of furious flaps,
and one of them suddenly gave way about 2 ft. from its
base, upon which the other one failed also. The inventor
stated that he was then engaged in the construction of a
larger machine on the same principle, but since then

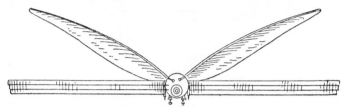

FIG. 14.—KAUFMANN—1867.

nothing more has been heard of it. He proposed to secure
stability by letting down or raising up a long " pendule"
with telescopic joints, so as to adjust the center of gravity
and keep the machine in a horizontal position, but it may
well be doubted whether this would have proved effective.

At a meeting of the British Aeronautical Society, in
1871, Mr. *R. C. Jay* exhibited a model to illustrate a
method which he proposed in order to use wings of any
length and weight without loss of power. This consisted
of two pairs of oscillating wings moving on the same
shaft. It was expected that the forces generated by their
motion would hold the machine is equilibrium, and that
one pair of wings would be aided by the current of air,
or whirlpool, produced by the movement of the other
pair. This does not seem to have answered, for in 1877
the same inventor presented a model to the same society,
illustrating a method of obtaining a figure of 8 or sculling
action with one pair of wings, but at the same time
Mr. *Jay* candidly stated that "although he had made a
great many experiments, he had not yet succeeded in
making a propeller (wings) sufficiently simple and effective
for practical purposes."

It is said that about the same time an *optician* of Leipsic
made a small steam bird, mounted on a globular steam
boiler and actuated by a cylinder of 2 in. stroke, working
wings 32 in. long. This machine would rise vertically
3 ft., the wings making about three beats during the
flight, but the boiler limited the performance. It contained spirits of wine only sufficient for 38 seconds, and
the apparatus was but a toy.

In 1871 *Prigent* designed the apparatus shown in fig. 15,

which was evidently suggested by the dragon-fly ; this is a favorite idea with aviators, who, as we have seen already, have proposed the combination of two pairs of wings over and over again. It was intended to be driven by steam, but although in that same year *Moy* had produced a steam-engine and boiler weighing but 27 lbs. per horse power, and *Stringfellow*, in 1868, has shown one claimed to weigh but 13 lbs. per horse power (both applied to aeroplanes), no attempt seems to have been made to experiment with *Prigent's* device. The fact is, that even the weight of the engines mentioned was too great, for it did not include the fuel and water, which for a non-con-

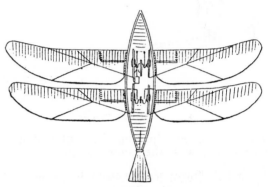

FIG. 15.—PRIGENT—1871.

densing steam-engine would amount to about 26 lbs. more horse power *per hour*, and this did not compare favorably with the motive power of birds. The pigeon, for instance, is known, both by dynamometric experiment and computation, to develop in ordinary flight from 160 to 425 foot-pounds of energy per minute for each pound of his weight, and as his pectoral muscles, which constitute *his* engine, generally compose $\frac{10}{43}$ of his weight, we have for the weight of *his* motor from

$$\frac{33.000 \times 10}{425 \times 43} = 18 \text{ lbs. to } \frac{33.000 \times 10}{160 \times 43} = 48 \text{ lbs.}$$

per horse power developed, including the fuel which enables him to fly for 10 to 12 hours at a stretch.

Hopeless, therefore, of accomplishing anything practical with steam-engines, experimenters with wings next turned their attention to springs or reservoirs of energy of various kinds, and with these they have succeeded in devising a number of toys which fly creditably for a few seconds. Clock springs were first tried, but they were found to be unduly heavy, and in 1871 *Jobert* brought out his first mechanical bird, shown in fig. 16, driven by india-rubber in tension. The wings were arranged so as to change their plane automatically while flapping, in order to imitate the flexions of the natural wings, and the equilib-

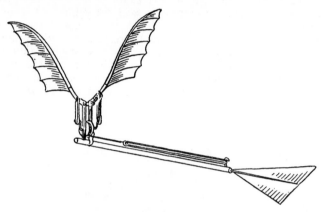

FIG. 16.—JOBERT—1871.

rium was secured by adjusting the center of gravity so as to correspond with the center of pressure due to the angle of flight.

In 1872 *Pénaud*, who had already succeeded (1870 and 1871) in compassing flight with the superposed screws and

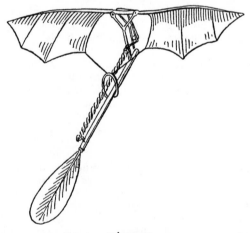

FIG. 17.—PÉNAUD—1872.

with the aeroplane, which will be noticed hereafter, by the force of twisted rubber, applied the same motor to a mechanical bird, which is shown in fig. 17. The wings beat straight down, and the propulsion is obtained from the flexion of their outer edges produced by the reaction of the air. The bird is unable to rise from the ground, but upon being thrown off the hand it first descends some 2 ft., and then, having acquired the initial velocity needed for support, it flies for a distance of 50 ft. in 7 seconds, rising at the same time about 8 or 9 ft. above the point of departure, the equilibrium being perfectly maintained by the tail.

Simultaneously with this M. *Hureau de Villeneuve*, the permanent Secretary of the French Aeronautical Society, brought out his mechanical bird, which is shown in fig. 18. In this the plane of the wings is inclined at an angle of 45°, and the power is obtained from twisted rubber. In consequence of the peculiar motion of the wings, this model was able to start direct from the ground, but owing

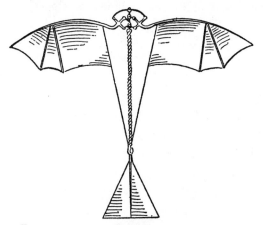

FIG. 18.—HUREAU DE VILLENEUVE—1872.

to the limited power of the rubber spring it only rose to the height of 4 ft., and then descended, forming a parachute. It was subsequently modified so that it would fly horizontally for a distance of 24 ft., at a velocity of 20 miles per hour.

M. *De Villeneuve* has been promoting aviation by flapping wings for the past 25 years. He has, first and last, designed something like 300 experimental models, so that his garret is a complete aviary of artificial birds. He built, some years ago, a huge steam bird on the model of a bat. Being aware that there was at that time no sufficiently light and reliable steam-engine with its boiler to furnish the power required, he placed only the engine on

the bird, and connected it by a hose with a boiler on the
ground. Upon trial, as soon as the steam was turned on
the wings beat violently, and the apparatus rose with the
inventor aboard. He grew nervous for fear that he would
get beyond the length of his hose, and shut off steam suddenly, upon which the bird fell and smashed one of its
wings. It is still in existence, and the inventor is awaiting the development of a very light motor in order to
resume his experiments with this great bird, which is
some 50 ft. across.

In 1872 M. *Jobert* brought out his second mechanical
bird, shown in fig. 19. This is driven by twisted rubber,
as being more manageable than rubber in tension, and
consists of four wings beating alternately in pairs—as a
horse trots---in order to produce continuous and uniform
support and equilibrium, instead of the jerking motion
observable in other apparatus. This flew fairly well, but
a measurement of the foot-pounds developed and of the
results obtained, in this as well as in the three other
mechanical birds previously described, led to the inference
that there was great waste of power, as compared with
that of birds. This was attributed to the rigidity of the

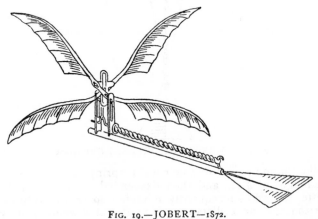

FIG. 19.—JOBERT—1872.

front edge of the wings in all these models, and accordingly in 1876 *Tatin* took the problem up again and succeeded, by a double eccentric working two levers connected to the front edge of the wing, in giving it a twisting motion similar to that of the bird. His apparatus
flew some 65 ft., with rather less weight of rubber.

In 1889 *Pichancourt* carried the matter still further in
the mechanical bird shown in fig. 20, in which there is a

FIG. 20.—PICHANCOURT—1889.

triple eccentric, each one actuating a lever fastened to a different point in the wings. His larger models, measuring 17½ in. from tip to tip of wings, and weighing 1½ oz., are said to have flown up to a height of 25 ft. and to a distance of 70 ft. against a slightly adverse wind.

Now here are no less than six artificial birds, each with a somewhat different wing-motion, and they all fly, when urged by the energy stored in twisted rubber. The question, therefore, occurs why practicable machines, to carry passengers, cannot be built by substituting some prime mover for the rubber; and the answer is that all these models are so wasteful of power that there is no motor known sufficiently light, in proportion to its energy, to take the place of the rubber. The best that seems to have been done with the latter was to obtain a flight of 7 seconds with flapping wings, and with the expenditure of energy at the rate of 600 foot-pounds per pound of twisted rubber. As there are 550 foot-pounds per second in a horse power, a primary motor, *with its supplies*, should in the same proportion weigh no more than :

$$\frac{600}{7 \times 550} = 6.4 \text{ lbs. per horse power,}$$

and there are none such known in practical operation.

Undeterred by this disheartening fact, M. *De Louvrié* designed, in 1877, the apparatus shown in fig. 21, which he calls the " Anthropornis," and which consists of a pair of wings, resembling those of the swallow, fastened to a hull mounted upon wheels, and intended to be actuated by a steam-engine or a petroleum motor. A spring is to contribute to the downward stroke, and is to be raised by the motor on the up stroke. M. *De Louvrié* is a veteran

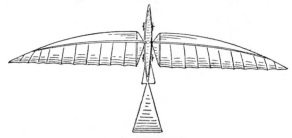

FIG. 21.—DE LOUVRIÉ—1877.

in promoting aviation, and his writings show a better understanding and firmer grasp of the question than most of those which have been published on this intricate subject. He had proposed, in 1863, a sort of kite-like flying machine, which will be noticed under the head of Aeroplanes, and it is said that, in 1888, he presented his latest views before a commission of the French Academy of Sciences, supplementing them with certain experiments, from which he drew the conclusion that an apparatus capable of carrying four passengers needed no more than 3 horse power to drive it at the rate of 67 miles per hour. It may be inferred that the French Commission was not convinced, from the fact that no action has been taken upon the proposal.

In the United States very few experiments seem to have been made with flapping wings, and no records of them are attainable. Investigation is, therefore, limited to such proposals as have been patented, and it is found that, aside from balloons, less than 30 flying machine patents have been taken out, of which four are for flapping wings.

The first of these, in order of date, seems to have been the proposal of Mr. *W. F. Quinby*, who patented, in 1869, an apparatus to be operated by man power, consisting of a pair of side wings and a tail, all to be flapped by a series of cords attached to the operator, who is encased in a cuirass which maintains the wings at about the height of his waist. The surfaces shown in the drawings are quite insufficient to sustain the weight, and in 1872 Mr. *Quinby* took out another patent for a modification of his apparatus, in which he added dorsal surfaces, so that the wings and the tail were continuous and resembled the supporting surfaces of a bat.   The arrangement for imparting motion was ingenious but futile, because of the inefficiency of muscular power, which has already been stated.

In 1876 Mr. *F. X. Lamboley* patented a framework shaped like the wings of a bird, and covered with a wire netting to which birds' feathers were fastened so as to give a valvular action.   Human power was relied upon to impart motion through a trapeze-bar and platform, and of course it would prove inadequate.

In 1877 Mr. *M. H. Murrell* patented an apparatus consisting of a pair of pivoted side wings and a tail, to be also operated by man power.   The wings were furnished with slats similar to those of a Venetian blind to close on the down stroke, and open when going up.   An investigation of what others had attempted would probably have saved the inventor some misspent time and ingenuity.

So little has been effected with flapping wings that a number of American inventors seem to have turned their attention to various arrangements of revolving vanes. Of these *A. P. Keith* patented, in 1870, an aerial car with paddle-wheels revolving in a transverse plane, for the purpose of lifting and propelling.   *Thomas Green* patented, in 1873, an apparatus with two wheels, each with four revolving blades passing through the air flatwise on the down stroke and edgewise on the up, and *M. H. Baldwin* patented, in 1890, an aerial vessel in which weight is to be supported by a set of wheels containing feathering vanes ;   the wheels revolving in opposite directions on longitudinal shafts.   All of these are worthless, as is also the patent of *I. M. Wheeler* of 1887, which covers the arrangement of a number of oscillating frames superposed to each other on a mast, and carrying slats similar to those of a Venetian blind ;   these various devices only being mentioned to illustrate how ingenuity has been wasted upon mechanical details, while scarcely any attention seems to have been given to the devising of the lightest possible motive power.   Each fresh inventor of winged machines is apt to imagine that his predecessors did not succeed because they did not hit upon the right method of imitating the complicated and swift motions of the birds.   Thus Mr. *H. Sutton*, of Australia, communicated to the British Aeronautical Society, in 1888, that experiment and observation had convinced him that the tips of the bird's wings describe, when viewed from the side, the outline of an inverted cone with rounded base, instead of the figure of 8 motion described by Dr. Pettigrew and

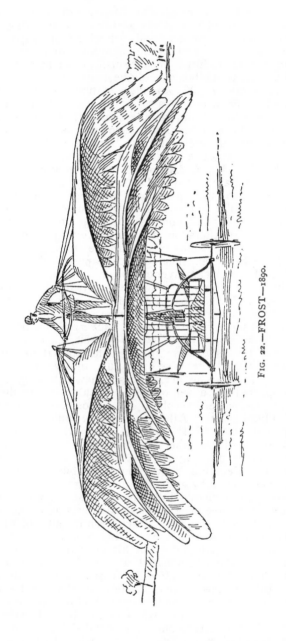

FIG. 22.—FROST—1890.

Professor Marcy. He had accordingly made a model, driven by clockwork, to test the truth of his theory. This model was not capable of free flight (steel springs and clockwork being much heavier than rubber, in proportion to their stored energy), but when suspended at the end of a counterweighted lever, resting upon an upright support with a ball-and-socket joint, it flew in a circumference of about 12 ft. by the flapping action of the wings. By slightly modifying the stroke of either wing it was made to fly from right to left or from left to right. By altering the guide-rods, which governed the direction of the stroke, it could be made to fly upward at any desired angle, but the important, the vital question of an efficient motor was left untouched by the inventor.

Still, earnest attempts are occasionally made in the direction of light motors. At the meeting of the British Aeronautical Society, in 1890, a photograph was shown of a steam-bird machine, designed and built by Mr. *E. P. Frost*, which is represented in fig. 22. The wings, which are 30 ft. from tip to tip, are in exact imitation of those of the crow, and the various positions which they assume during a stroke are shown in the picture. The weight of the machine, including engine and boiler, is about 650 lbs. It was expected to carry in addition the weight of a man in the air, but it was said that the maker of the engine failed in his contract to secure the necessary power, and the apparatus did not fly.

At the same meeting Mr. *H. Middleton*, who has been advocating for several years winged apparatus as superior, in his judgment, to aeroplanes, exhibited two bird machines, one weighing 20 lbs., with a wing-spread of nearly 12 ft., and the other of between 10 and 12 lbs. weight with a wing-spread of some 9 ft. He also showed an aeroplane weighing somewhat over 20 lbs., with sustaining planes of 14 ft. across and a screw of 4 ft. diameter, in order to compare its performance with those of the bird machines. Only the smaller of these latter was shown in action, but its balance was not properly adjusted, and although it raised itself from the sustaining horizontal rope during the first few strokes, it soon rested again upon the rope, and on the pressure being raised during a subsequent run, the right wing broke and terminated the experiment.

The aeroplane, being similarly suspended, moved along the rope at a moderately good pace, but without raising itself on the air, and that experiment was brought to an untimely end by the rupture of a joint on the propeller shaft.

Probably the most original conception ever presented

for a flying machine is that of M. *G. Trouvé*, who has just revived (1891) the proposal for his mechanical bird, which was first presented to the French Academy of

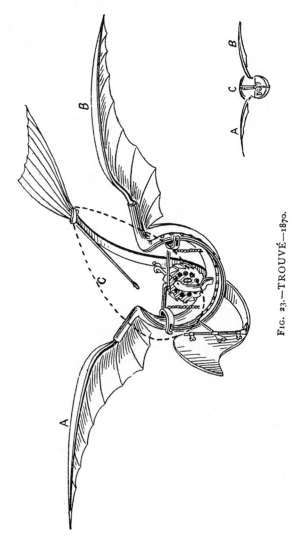

FIG. 23.—TROUVÉ—1870.

Sciences in 1870. This is shown in fig. 23, and consists of two wings, *A* and *B*, connected together by a " Bourdon" bent tube, such as is used in steam gauges. The peculiarity of this tube, as is well known, is that as pres-

sure increases within it the outer ends move apart, and
as pressure diminishes they return toward each other.
M. *Trouvé* increases the efficiency of this action by placing a second tube within the first, and in the experimental
model he produces a series of alternate compressions and
expansions by exploding 12 cartridges contained in the
revolver barrel *D*, which communicates with the tube.
This produces a series of energetic wing strokes which
propel and sustain the bird in the air in connection with
a silk sustaining plane indicated at *C.*

The manner of starting the bird is equally ingenious
and peculiar, and is shown in fig. 24.   The bird is suspended from a frame by a thread, which, being attached
to the hammer, keeps the latter off the cap.   A second
thread holds the bird back from the perpendicular, while
a common candle *A* and a blow-pipe flame *B* complete the
preparations.   Upon the thread being burned at *A* the bird
swings forward from position 1 to position 2, where the
suspending thread is burned by the blow-pipe *B*, the
hammer falls on the cap, an explosion ensues, the wings
strike downward violently, and the bird flies on an upward
course, as shown in position 3.   Then the gases escape
from the Bourdon tube, this recovers its shape, raises the
wings and actuates two pawls which rotate the revolver
barrel and work the hammer, so that a fresh explosion

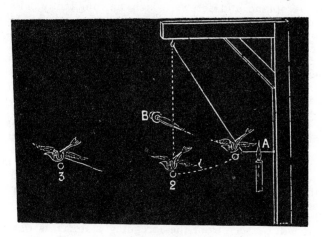

FIG. 24.—STARTING TROUVÉ'S BIRD.

occurs and the bird continues to fly.   When the 12
cartridges are exhausted the bird glides gently to the
ground, being sustained by its wings and aeroplane as by
a parachute.   It has thus flown 75 to 80 yards.

This motor is evidently very simple and very light, and for a practical flying machine M. *Trouvé* proposes to substitute for the cartridges a supply of compressed hydrogen gas, which, when mixed with about three parts of air, becomes an explosive mixture to be fired by the electric spark. Thus the motor would derive the greater portion of its power direct from the atmosphere as wanted, there would be no danger of premature explosion as with fulminates, and, hydrogen being only $\frac{1}{14}$ the weight of air, the weight and the equilibrium of the apparatus would vary but little when supplies became exhausted. Moreover, it is probable that no cooling agent would be required, as in ordinary gas engines, because the tube exposes so great a surface that it is to be expected that the heat would pass into the air while under motion, and that, as there is no piston to be lubricated, a moderate heating would not prove objectionable.

M. *Trouvé* started out with the assumption that a motor for aerial navigation should not weigh over 8 lbs. to the horse power. He presented to the French Academy of Sciences, in 1886, an electric motor, weighing but 7.7 lbs. per horse power, working an aerial screw, which will be more fully noticed when that subject is treated, and on August 24 last (1891) he deposited with the same body a sealed letter containing the drawings and description of an aeroplane and screws, which, he confidently believes, provide a final solution for the problem of aerial navigation, and which will also be noticed under the head of Aeroplanes.

Meanwhile other inventors are also working in the same field, and the English papers have contained sundry paragraphs, within the last few months, concerning a flying machine some 45 ft. across, in the form of a bat, which is being built in Coventry for Major Moore, of India. It is to be driven by an electric motor, but the descriptions do not make it clear whether this is to be by beating wings or with fixed wings, as in an aeroplane. The cost incurred is stated at over $5000, and the trial is to take place at the Crystal Palace. Should this take place in time, it will be noticed, as well as the great apparatus now being completed by Mr. Maxim, when we come to discuss aeroplanes.

It will be seen from the foregoing statements of what has been accomplished with beating wings, that the principal questions are those of motive power and of proportion of surfaces to weight, and the reader will probably first inquire as to what is really the power developed by birds in their flight. The answer must unfortunately be, that it is not accurately known. A great many computations have been made, based upon more or less plausible assumptions, but none of these computations can be absolutely accepted as correctly based upon indisputably measured data.

This ceases to be surprising when we consider that there is no creature so willful, so swift, and so easily affrighted as the bird, and that once in the air, he will not lend himself to be measured experimentally. Mathematicians have, therefore, partly resorted to conjectures for their data. Thus Napier assumed that a swallow weighing 0.58 oz. must beat his wings 2,100 times per minute while going $33\frac{1}{3}$ miles per hour, in order to progress and sustain his weight, and that it therefore expended $\frac{1}{13}$ of a horse power. In point of fact, the bird only beats about 360 times a minute, and is chiefly sustained by the vertical component of the air pressure on the under side of the wings and body, due to the speed, instead of by the direct blow of the wings downward, as supposed in the orthogonal theory already alluded to.

Other mathematicians, starting from the fact that a weight falls about 16 ft. during the first second, and in so dropping does work, have assumed that a bird in horizontal flight, being then sustained, performs a certain fraction of this work. It is evident, however, that if the bird does not drop, the fraction assumed is purely arbitrary, and that such calculations must be quite worthless.

Experiments to measure directly the power expended have proved failures, and resort has been had to indirect measurements.

Thus Dr. W. Smyth, of Edinburgh, succeeded in measuring with a dynamometer the strain exerted by a 12-oz. pigeon while flexing its wings, when excited by a current of electricity, and found it capable of raising 120 lbs. one foot high in a minute, or at the rate of 160 foot-pounds per pound of bird. Professor Marey performed the same experiment on the buzzard and on the pigeon, and ascertained the contractile strength of their muscles to be 18.46 and 19.91 lbs. to the square inch respectively ; * but as he was unable to measure satisfactorily the rapidity with which the muscles contracted, he did not calculate the foot-pounds.

Mr. Alexander, starting with the assumption that a 2-lb. pigeon makes 180 completed strokes per minute, each stroke with an amplitude of 1.5 ft. at the center of pressure, calculates the power exerted as being $2 \times 180 \times 1.5 = 540$ foot-pounds per minute, or at the rate of 270 foot-pounds per pound of bird. This is plausible ; but the

---

* " Vol des oiseaux," page 92.

most satisfactory computations are those made by Pénaud
from observations of the direct velocity of ascent of various
birds. From these he concludes that the pigeon, for instance, expends for rising 579 foot-pounds per minute, and
that the proportion of horse power to weight is as follows :

For the peacock, one horse power for every 66 lbs.
"   "   pigeon,   "   "   "   "   "   57   "
"   "   sparrow,   "   "   "   "   "   48½   "
"   "   sea pie,   "   "   "   "   "   26   "

This, however, is merely the work of elevation, such as
would be performed upon a solid support, in addition to
which the bird has to overcome the resistance of the air to
his motion, and to derive support from this mobile fluid.
Pénaud calculates that this additional work amounts to
over 1,000 foot-pounds per minute, so that the total work
done by the pigeon in rising to a perch 35 ft. above the
ground amounts to 1,650 foot-pounds per minute, or 1
horse power for every 20 lbs. Moreover, it must be remembered that the pectoral muscles of birds, which constitute
their motor, comprise but one-quarter to one-sixth of their
total weight, so that in this particular case the relative
weight of the motor is only about 5 lbs. per horse power
for the force exerted in rising.

These are formidable figures, but they cease to be discouraging when we reflect that the effort of rising is evidently a maximum, and that birds seldom perform it in a
nearly vertical direction except for short distances, and that
the exertion is clearly so severe that the feat is usually performed only by the smaller birds, which, as previously explained, must possess greater energy in proportion to their
weight than those exceeding a few ounces. Heavy birds
can only rise at angles less than 45°, and even then they
exert for a short time far more than their mean strength,
the latter being, for all animals, only a fraction of the
maximum possible effort. Thus man, who is usually estimated as capable of exerting 0.13 horse power for 10
hours, can develop 0.55 horse power for 2½ minutes, and
nearly a full horse power for 3 or 4 seconds ; and it seems
probable that similar proportions obtain for birds, the
emergency effort being three or four times the average
performance, and the possible maximum about twice as
great as the emergency effort.

Pénaud states that the ring-dove dispenses in full flight
217 foot-pounds per minute ; but he does not give figures for this, so that they can be checked. Goupil estimates the work done by a pigeon weighing 0 925 lbs. at
1,085 foot-pounds per minute in hovering and 119 foot-
pounds per minute in flight ; but the latter is arrived at by

reasoning from analogy. It is evident that the power exerted in horizontal flight is much less than that required for rising or for hovering ; but until a bird is taught to tow behind him some dynanometric arrangement at a regular rate of speed, and on a level course, it will be difficult to settle exactly what are the feet-pounds expended in ordinary performance.

In 1889 Captain de Labouret, an expert in the solution of ballistic problems, analyzed mathematically two series of photographs of a gull weighing 1.37 lbs., and just starting out in flight with 5 wing beats per second, as obtained by by Professor Marey with the chrono-photographic process. The calculations showed that the bird expended in this act an average of 3,152 foot-pounds per minute, or 2.303 foot-pounds per pound of his weight ; and as Professor Marey shows that from his other observations of the reduced amplitude and rapidity of the wing beats, the same bird does only expend in full flight $\frac{1}{5}$ of the effort required at starting, the conclusion may be drawn that the gull in full flight expends some 460 foot-pounds per minute for each pound of his weight.

This estimate seems plausible to me, and agrees with my own figures, but it is not accepted by all aviators. The *Revue Scientifique* of November 28, 1891, contains two articles disputing the conclusions—one by Mr. V. Tatin, an expert aviator, who claims that the accelerations of the bird have been erroneously calculated ; that the center of pressure under the wing is $\frac{4}{7}$ of the distance from its root instead of $\frac{2}{3}$, as usually assumed, and who figures out from the velocity of this new center of pressure, and from the known trajectory that the bird in full flight only expends from 33 to 197 foot-pounds per minute for each pound of his weight.

The second article is by Mr. C. Richet, the editor of the *Revue Scientifique*, who, having ascertained the volume of carbonic acid exhaled by a bird at rest, assumes, from experiments on other animals, that in full flight he will give out three times as much, and that the difference represents an effort of 105 foot-pounds per minute per pound of bird.

These two articles, being the most recent computations by earnest students of the subject, are here mentioned chiefly to illustrate how greatly aviators vary in estimates of the power expended, and how many elements have to be assumed in making such computations.

In the absence of direct measurements, and of positively satisfactory computation by others, of the feet-pounds expended in horizontal flight, I believe that an approximation may be obtained by analyzing and calculating the various elements which combine to make up the aggregate of the resistance to forward motion in horizontal progression ; and as this method promises to be useful in computing the power required by artificial flying machines, I venture to set it out at some length, applying it to the domestic pigeon as being more convenient to compare with the results of the calculations of others. For this purpose two dead pigeons were selected, weighing as near as practicable 1 lb. each, and their dimensions were accurately measured as follows :

CROSS SECTION AND HORIZONTAL PROJECTION OF PIGEONS.

	Pigeon No. 1.	Pigeon No. 2.
Largest cross section of body..............	4.9 sq. in.	5.3 sq. in.
" " " edge of wings.....	5.02 " "	4.88 " "
Weight of bird, freshly killed	1 lb.	0.969 lb.
Horizontal area of both spread wings......	90.35 sq. in.	99.86 sq. in.
" " " body projected....... ...	22.49 " "	24.01 " "
" " " tail spread...............	19.72 " "	27.17 " "
	132.56 sq. in.	151.04 sq. in.

These dimensions all require the application of coefficients in calculating their action upon the air. Thus the wings are concave, and give a greater sustaining power per square foot than a flat plane ; the body is convex, and affords less than a plane, while the tail is slightly concave, but partly ineffective from its position. Previous experiments have indicated that, in the aggregate, the supporting power is about 30 per cent. more than that of a flat plane of equal area, so that [in the calculations which follow the supporting surfaces will be assumed at 1.3 sq. ft. to the pound instead of the 1 square foot to the pound which the average of the measurements seems to indicate.

It will be remembered that experiments with parachutes indicate a coefficient of resistance of 0.768 for the convex side and of 1.936 for the concave side, as compared with the plane of greatest cross section.

The cross sectional area of the body is assumed at 5 square inches or 0.03472 of a square foot, and to this a coefficient is applied of one-twentieth of a flat plane, or 0.05, in consequence of its elongated, fusiform shape. This agrees well with experiments on the hulls of ships of "fair" shape.

The cross sectional area of the wings is also taken at 5 square inches, or 0.03472 of a square foot ; but the coefficient here assumed is about one-seventh, or 0.15, in consequence of its shape, which is ogival, or rather something like only half of a Gothic arch.

The friction of the air is omitted, as being entirely too small to affect the results in a case where so many co-efficients have to be approximated.

The angle of flight is ascertained by selecting from the table previously given of air reactions, the coefficient which will give the nearest approximation to a sustaining "lift" to support the weight, and from this angle the "drift" is obtained to calculate the resistance of the surface.

The velocity V is in feet per minute, and the pressure P on a plane at right angles to the current by the Smeaton formula is in pounds per square foot. The following are the calculations:

20 miles per hour—$V = 1760$ ft. $P = 2$ lbs.

Lift, 12°, $1.3 \times 2 \times 0.39 = 1.014$ lbs. sustained.

		Resistance.	Power.
Drift, 12°,	$1.3 \times 2 \times 0.0828 = 0.21520$ lb.	$\times 1760 =$	378.7 ft. lbs.
Body resistance,	$0.03472 \times 2 \times 0.05 = 0.00347$ "	$\times 1760 =$	6.1 "
Edge wings,	$0.03472 \times 2 \times 0.15 = 0.01040$ "	$\times 1760 =$	18.3 "

0.22907 lb.   403.1 ft. lbs.

30 miles per hour—$V = 2640$ ft. $P = 4.5$ lbs.

Lift, 5°, $1.3 \times 4.5 \times 0.173 = 1.012$ lbs. sustained.

		Resistance.	Power.
Drift, 5°,	$1.3 \times 4.5 \times 0.0152 = 0.08892$ lb.	$\times 2640 =$	234.7 ft. lbs.
Body resistance,	$0.03472 \times 4.5 \times 0.05 = 0.00781$ "	$\times 2640 =$	20.6 "
Edge wings,	$0.03472 \times 4.5 \times 0.15 = 0.02343$ "	$\times 2640 =$	61.9 "

0.12016 lb.   317.2 ft. lbs.

40 miles per hour—$V = 3520$ ft. $P = 8$ lbs.

Lift, 3°, $1.3 \times 8 \times 0.104 =$ 1.082 lbs. sustained.

		Resistance.	Power.
Drift, 3°,	$1.3 \times 8 \times 0.00543 = 0.05647$ lb.	$\times 3520 =$	198.7 ft. lbs.
Body resistance,	$0.03472 \times 8 \times 0.05 = 0.01389$ "	$\times 3520 =$	48.9 "
Edge wings,	$0.03472 \times 8 \times 0.15 = 0.04166$ "	$\times 3520 =$	146.6 "

0.11202 lb.   394.2 ft. lbs.

50 miles per hour—$V = 4400$ ft. $P = 12.5$ lbs.

Lift, 2°, $1.3 \times 12.5 \times 0.07 =$ 1.137 lbs. sustained.

		Resistance.	Power.
Drift, 2°,	$1.3 \times 12.5 \times 0.00244 = 0.03965$ lb.	$\times 4400 =$	174.5 ft. lbs.
Body resistance,	$0.03472 \times 12.5 \times 0.05 = 0.02170$ "	$\times 4400 =$	95.5 "
Edge wings,	$0.03472 \times 12.5 \times 0.15 = 0.06510$ "	$\times 4400 =$	286.5 "

0.12645 lb.   556.5 ft. lbs.

60 miles per hour—$V = 5280$ ft. $P = 18$ lbs.

Lift, 1½°, $1.3 \times 18 \times 0.052 =$ 1.217 lbs. sustained.

	Resistance.	Power.
Drift, 1½°,	$1.3 \times 18 \times 0.00136 = 0.0318$ lb.	$\times 5280 = 167.9$ ft. lbs.
Body resistance,	$0.03472 \times 18 \times 0.05 = 0.0312$ "	$\times 5280 = 164.7$ "
Edge wings,	$0.03472 \times 18 \times 0.15 = 0.0937$ "	$\times 5280 = 494.7$ "
	0.1567 lb.	827.3 ft. lbs.

These figures are probably somewhat in excess of the real facts in consequence of the adoption of slightly excessive coefficients for the resistance of the body and wing edges, which coefficients in full flight may be as much as one-third less than those which have been estimated.

It will be noticed that, as the velocity and the consequent air pressures increase, the angle of incidence required to obtain a sustaining reaction or "lift" diminishes, and so does, therefore, the "drift" or horizontal component of the normal pressure, while the "hull resistance," consisting of that of the body and edges of the wings, is at the same time increasing. There will therefore be some angle at which these various factors will so combine as to give a minimum of resistance, and this is probably for most birds at an angle of about 3°, which in the case of our calculated pigeon requires a speed of 40 miles per hour in order to sustain the weight.

This angle of minimum resistance depends upon the relative proportions of the bird—i.e., upon the ratio between his surface in square feet per pound of weight, and the cross section of his body and wings, as well as their coefficient of resistance; and so, while the angle may not vary greatly, it needs to be ascertained for each case. Mr. Drzewiecki has calculated that for an aeroplane exposing a cross sectional area of 1 per cent. of its sustaining area (instead of the 7 per cent. which the measurements show for the pigeon), the angle of minimum resistance would be 1° 50′ 45″, and that it would be the same for all velocities. It does not follow, however, that the minimum of power required will coincide with the minimum of resistance, for the latter increases as the square, while the power grows as the cube of the speed. The calculations, therefore, show that the minimum of resistance occurs at 40 miles per hour, while the minimum of work done in foot-pounds is found at 30 miles per hour, and these two favorable speeds are about those observed from railway trains, as habitually practised by the domestic pigeon.

The estimates of the feet-pounds per minute indicate that the bird finds it less fatiguing to fly at 30 miles per hour than at 20; that his exertions are not much greater at 40 miles per hour, but that at 50 miles per hour he is expending rather more than his mean strength—the latter being probably about 425 foot-pounds per minute, nearly an average of the first four calculations, or about one-quarter of the maximum work done in rising, as estimated by Pénaud.

A flight of 60 miles within the hour is probably a severe exertion for the domestic pigeon, while the finer lines and greater endurance of the carrier pigeon enable him to maintain this speed for hours at a time ; but there is reason to believe that this must be nearly the limit of his strength, and that homing birds who have made records of 70 and 75 miles per hour were materially aided by the wind.

The calculations therefore appear plausible, and to agree fairly well with the estimates arrived at with different methods by others. They indicate that if a flying machine can be built to be as efficient as the domestic pigeon, its motor should develop one horse power for each 18 lbs. of its weight, provided it can give out momentarily about four times its normal energy, or that special devices, such as that of running down an incline or utilizing the wind, or some other contrivance are adopted to give it as tart and to enable it to rise upon the air.

The next question which the reader will probably want to ask, is as to the amount of supporting surfaces possessed by birds in proportion to their weight. Upon this point a good deal of information has been published ; and in 1865 Mr. De Lucy greatly cheered aviators by publishing a paper in which he showed that the wing areas of flying animals diminish as the weight increases, from some 49 square feet to the pound in the gnat to 0.44 square feet to the pound in the Australian crane ; and from which tables he inferred the broad law that the greater the weight and size of the volant animal, the less relative wing surface it required.

As thus stated, the assertion is misleading. For inasmuch as the supporting surfaces will increase as the square, and the weight will grow as the cube of the homologous dimensions, it was to be expected that wing surfaces would not increase in the same ratio as the weight if the strength of the parts remained the same ; and in 1869 Hartings published some tables of birds, in which he compared the square root of the wing surface with the cube root of the weight, and showed that their ratio became what he considered a somewhat irregular constant. Subsequent measurements and tables by Professor Marey have shown that this statement of Hartings is also slightly misleading, inasmuch as the so-called constant varies from 1.69 to 3.13, so that no broad law can be laid down as to any fixed relation between the surfaces and weight of birds of various sizes. The fact seems to be that while their structures are governed by the laws which limit the strength of materials (bones, muscles, feathers, etc.), yet there are differences in the resulting stresses, and in the consequent efficiency of the birds themselves, who are thereby led to adopt slightly different modes of flight ; and in 1884 Müllenhoff published an able paper, in which he divided flying animals into six series, in accordance with the ratio between their weight and their wing surface, as well as their methods of flight. As the tables of De Lucy, Hartings, Marey and Müllenhoff are all easily accessible in print, they will not be repeated here ; but the following table is considered more valuable than any of them. It has been compiled from " L'Empire de l'air" of Mr. Mouillard, a very remarkable book, published in 1881, which contains descriptions of the flight of many birds and accurate measurements of their surfaces and weights.

TABLE OF SUPPORTING AREAS OF BIRDS.
MEASURED BY L. P. MOUILLARD.
COMPILED BY S. DRZEWIECKI.

Scientific Name.	Common Name.	Sq. Ft. per Lb.	Lbs. per Sq. Ft.	Corr'sp'ding speed for a plane at 30 Miles per hr
Nyctinomous ægypticus	Bat....	.64	0.131	15.9
Upupa epops...............	Peewit.	3.62	0.276	23.1
Cotile rupestris.............	Swallow.........	3.62	0.276	23.1
Budytes flava	Wagtail	3.49	0.286	23.5
Galerita cristata I..........	Lark............	3.18	0.315	24.6
Caprimulgus	Goatsucker.....	3.17	0.314	24.6
Galerita cristata II..........	Lark............	3.06	0.327	25.1
Accipter nisus..............	Sparrow-hawk...	3.00	0.333	25.3
Pteropus Geoffroyi..........	Bat..............	2.79	0.362	26.2
Coracias garrulus...........	Roller..........	2.76	0.363	26.5
Tringa canutus.........	Knot.	2.64	0.380	27.0
Falco tinnunculus...........	Falcon..........	2.48	0.403	27.9
Passer domesticus I.........	Sparrow........	2.42	0.414	28.2
Vanellus cristatus...........	Lapwing........	2.40	0.417	28 3
Passer domesticus II........	Sparrow...	2.36	0.424	28.6
Cypselus apus..............	Martinet........	2.35	0.426	28.6
Larus melanocephalus I. ...	Gull............	2.35	0.426	28.6
Glareola torquata...........	Glareola........	2.32	0.431	28.8
Larus melanocephalus II....	Gull	2.30	0.435	28.9

Scientific Name.	Common Name.	Sq. Ft. per Lb.	Lbs. per Sq. Ft.	Corr'sp'ding speed for a plane at 30 Miles per hr.
Turtur ægypticus........	Egyptian Dove.	2.27	0.441	29.2
Otus brachyotus........... ...	Owl............	2.26	0.443	29.2
Strix flammea..............	"	2.26	0.443	29.2
Milvus ægypticus......... .. .	Kite............	2.19	0.457	29.7
Petrocincla cyanea	Blackbird.......	2.18	0.460	29.7
Alcedo hispada I....	Kingfisher.	2.11	0.475	30.3
" " II.........	"	2.11	0.475	30.3
Buphus minutus........... ..	Crane...........	2.02	0.495	30.9
Scolopax gallinula I...	Snipe......... ..	1.96	0.510	31.4
Ephialtes zorca.............	Scops..........	1.90	0.526	31.8
Alcedo hispida III..........	Kingfisher......	1.87	0.535	32.1
Corvus ægypticus........	Rook...........	1.74	0.575	33.3
Astur palumbarius.........	Goss-hawk......	1.73	0.579	33.4
Ibis falcinellus..............	Ibis..	1.66	0.603	34.1
Sturnus vulgaris............	Starling........	1.65	0.606	34.2
Scolopax capiensis...	Snipe...........	1.65	0.606	34.2
Corvus corax..............	Raven..........	1.62	0.614	34.5
Scolopax gallinula II..	Snipe..........	1.60	0.625	34.7
Philomachus pugnax.... ..	Water-fowl......	1.48	0.634	36.1
Ardea nycticorax....	Night Heron....	1.43	0.700	36.7
Ciconia alba..............	Stork..........	1.40	0.715	37.1
Charadrius pluvialis........	Plover..........	1.38	0.725	37.4
Columbia ægyptica I........	Egyptian pigeon	1.37	0.730	37.5
Falco peregrinus............	Falcon..........	1.29	0.775	38.6
Rallus aquaticus............	Rail............	1.28	0.781	38.8
Pandion fluvialis..	Balbuzzard......	1.26	0.795	39.2
Neophron percnopterus....	Egypt'n vulture	1.18	0.848	40.4
Columba ægyptica..........	" pigeon.	1.13	0.885	41.3
Numenius arquatus..... ...	Coulis..........	1.11	0.901	41.7
Ortyx coturnix............	Quail..	1.08	0.927	42.3
Recurvirostra avocetta......	Avocetta........	1.05	0.954	42.8
Œdicnomus crepitans..	Plover..........	0.926	1.079	43.6
Anas querquedula..........	Duck...........	0.864	1.158	44.2
Puffinus Kulhi...	Shearwater......	0.853	1.170	44.5
Gallinula chloropus........	Water-hen.	0.765	1.307	50.3
Numenius arquatus.........	Curlew	0.761	1.312	50.3
Pelecanus anocrotales.... ..	Gray Pelican...	0.732	1.365	51.3
Gyps fulvus..............	Tawny Vulture.	0.679	1.473	53.3
Otogyps auricularis.........	Oricou.........	0.664	1.473	53.9
Pterocles exustus...........	Running Pigeon	0.664	1.508	53.9
Procellaria gigantea	Giant Petrel....	0.640	1.561	54.9
Anser sylvestris............	Wild Goose.....	0.586	1.708	57.4
Meleagris Gallopavo..	Turkey	0.523	1.910	60.6
Anas clypeata, female......	Duck	0.498	2.008	62.2
" " male........	" ....,	0.439	2.280	66.2

Mr. Mouillard adopted a more rational method than other observers. Instead of merely measuring the surface of the wings, he laid the bird upon its back on a sheet of paper, projected the entire outline, and then measured the total area from which it gains support. The compilation has been made by Mr. Drzeweicki for a paper presented to the International Aeronautical Congress at Paris in 1889, in which he states the general law more accurately than his predecessors, by calling attention to the fact that the ratio of weight to surface will vary somewhat with the structure of the bird, and that the result will be that those possessing the lesser proportionate surface must fly faster in order to obtain an adequate support at the same angle of incidence.

I have added the last column in the table, showing the speed required to sustain the weight of a flat plane loaded to the same proportion of weight to surface as the bird, at an angle of incidence of $3°$. This speed merely approximates to the real flight of the bird, because it takes no account of the concavity of the wings, which, as previously explained, increases the effective bearing surface of the animal ; but it would require experimenting with each and every bird tabulated in order to give the true and varying coefficients.

B.—SCREWS TO LIFT AND PROPEL.

In describing the various proposals and experiments which have been made to compass artificial flight by means of rotating screws, the latter will chiefly be considered as instruments from which to obtain support of a given weight in the air. There is no question that they can serve as propellers if the support be otherwise obtained ; nor that if a screw can lift and sustain its own prime motor, it can also be made to progress horizontally, either by inclining it at the proper angle or by adding a vertical screw.

It was to be expected that when inventors found how difficult it is to obtain a lifting effect from flapping wings, they should turn to aerial screws to sustain them in the air. Man has succeeded in out-traveling both land and marine animals by substituting rotary motion for the reciprocating action of their limbs : the locomotive far outstrips the horse, and the paddle-wheel and screw have, for large vessels, superseded the oar, so that it seems natural to expect that some rotating device shall be found the preferable propeller, should aerial navigation ever be accomplished.

It will be seen, from the accounts which follow, that the chief obstacle has hitherto been the lack of a sufficiently light motor in proportion to its energy ; but there has recently been such marked advance in this respect, that a partial success with screws is even now almost in sight.

Curiously enough, the Aerial Screw considerably antedates the marine screw, although, unlike the latter, it has not been brought into practical use. We have already seen that *Leonardo Da Vinci* experimented with paper screws, which mounted into the air, as early as A.D. 1500, and we may add that a sketch has been found in his note books for a proposed aerial screw machine 96 ft. in diameter to be built of iron and bamboo framework, covered with linen cloth thoroughly starched. He probably abandoned all idea of constructing it when his experiments with models showed the power that would be required.

A similar proposal was made by *Paucton*, a learned mathematician, in 1768, when, in a treatise upon the Archimedean screw, he described an apparatus which he called a " Pterophore," consisting of two aerial screws, one to sustain and the other to propel, attached to a light chair. A man seated in the chair was expected to rotate these screws by means of gearing, and so raise himself through the air.

The first practical experiment known, however, is that of M. *Launoy*, a naturalist, and M. *Bienvenu*, a mechanician, who jointly exhibited before the French Academy of

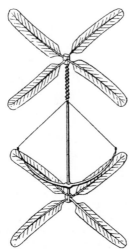

FIG. 25.—LAUNOY & BIENVENU—1784.

Sciences in 1784 the little apparatus shown in fig. 25. It consisted of two superposed screws, about one foot in diameter, each composed of four feathers inserted in sockets at the ends of a rotating axle. This axle was put into motion by the unwinding of a cord fastened to the two extremities of a bow ; and the report to the French Academy (May 1, 1784) says :

"The working of this machine is very simple. When the bow has been bent by winding the cord, and the axle placed in the desired direction of flight—say vertically, for instance—the machine is released. The unbending bow rotates rapidly, the upper wings one way and the lower wings the other way, these wings being arranged so that the horizontal percussions of the air neutralize each other, and the vertical percussions combine to raise the machine. It therefore rises and falls back afterward from its own weight."

Launoy & *Bienvenu* proposed also to build a large machine, and to go up in it themselves. It is not stated whether this was ever attempted ; but probably not, as a brief investigation must have satisfied them that they had no adequate primary motive power at hand to lift even its own weight in that way, and that with a secondary or stored power the machine would fly but for a few seconds.

Practically the same device was constructed by Sir *George Cayley* in 1795, and described by him in *Nicholson's Journal* for April, 1810 ; but whether he reinvented it or borrowed the idea from *Launoy* & *Bienvenu* is not stated. He mentions it merely as a toy, and his writings seem to indicate that he expected success to be achieved instead with an aeroplane to be driven by some sort of propelling apparatus, if only a sufficiently light first mover could be contrived.

Subsequently, *Deghen*, in 1816, *Sarti*, in 1823, and *Dubochet*, in 1834, all proposed and constructed models for flying machines on the vertical screw principle ; but they did not discover the necessary light motor to transform their models into practical machines.

In 1842 Mr. *Phillips*, the inventor of the "Fire Annihilator," succeeded in raising into the air an apparatus weighing in the aggregate 2 lbs., by means of revolving fans inclined about 20° from the horizontal. The motive power was evolved by the combustion of charcoal, nitre and gypsum making steam, as in the original fire annihilator ; and the engine consisted of rotating arms discharging steam direct into the atmosphere, and thus working by reaction, being the device known as the discovery of Hero, of Alexandria. Mr. *Phillips* exhibited a working model of his aerial machine at the Aeronautical Exhibition in London, in 1868 ; and in describing his experiment of 1842 he said :

" All being arranged, the steam was up in a few seconds, when the whole apparatus spun around like a top and mounted into the air faster than any bird ; to what height it ascended I had no means of ascertaining. The distance traveled was across two fields, where, after a long search, I found the machine minus the wings, which had been torn off from contact with the ground." This is undoubtedly the first machine which has risen into the air by steam power ; but the necessarily small capacity of the generator, and the wasteful though simple method of using the steam, limited its flight to a very few minutes, and removed it from the possibility of application on a practical scale.

In 1843 Mr. *Bourne*, the well-known English engineer, constructed some models of aerial screws, consisting of large fowl's feathers inserted in a cork, stuck on the top of a pine stick, to which a watch spring was attached, and succeeded in making them rise by the force of the coiled spring to the height of some 20 ft. ; but he recognized that the difficulty in the way of building a really navigable machine was to obtain " the right motive power."

This must also have been the conclusion of Mr. *Cossus*, who proposed, in 1845, the apparatus shown in fig. 26, which consists in three rotating aerial screws to be moved by steam power. The design is by no means devoid of

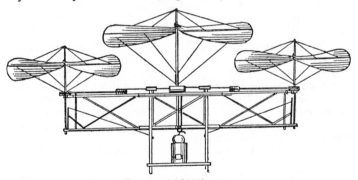

FIG. 26.—COSSUS—1845.

merit, for by hinging the outer and smaller screws, and varying their angle with respect to the machine, the latter can be made to travel in any direction desired, while sustained by the rotation of the middle screw. It cannot be learned that *Cossus* tried any practical experiments, for a simple inquiry into the weights and relative energy of the steam engines of his day and an investigation as to the power required to sustain his apparatus must have speedily convinced him that it had better be abandoned.

Analogous proposals were made in 1851 by Mr. *Aubaud*, who combined several screws with an aeroplane ; and by *Le Bris*, who designed a car surmounted by two screws turning in opposite directions, in order to overcome the tendency of the apparatus to rotate on its own axis, as the consequence of the horizontal component of the thrust of a single screw.

It was to overcome this same objection that, in 1859, Mr. *Bright* designed and patented the apparatus shown in fig. 27, the axles of the screws consisting of tubes, rotating in opposite directions, one inside of the other.

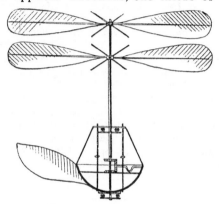

Fig. 27.—BRIGHT—1859.

Mr. *Bright* seems to have planned the machine to be suspended beneath a balloon, and to be worked by man power, in order to alter or to maintain the altitude at will, and thus save the expenditure of ballast in rising or of gas in descending. Its beneficial effects, however, seem to have proved so small—solely, it may be said, from the inadequacy of the motive power employed—that it has not come into practical use.

These various efforts were somewhat desultory, and not followed up by anything like scientific experiments ; but in 1863 there was in Paris a great " boom" in projects for navigating the air by means of aerial screws, and the French espoused its promotion with great enthusiasm. M. *de Ponton d'Amécourt* and M. *de la Landelle* had already studied the action of the screw upon the air, when, in July, 1863, M. *Nadar*, a prominent photographer, invited to his reception rooms the *élite* of the press, of science, and of artists, and treated them to a first reading of his famous " Manifesto upon Aerial Automotion," which appeared the next day in the press, and was republished and commented upon throughout the whole of Europe.

In this manifesto, written with much eloquence, *Nadar* expressed the opinion that the principal obstruction in the way of navigating the air was the attention which had been given to balloons ; that, in order to imitate nature, a flying machine must be made heavier than the air. Also that the surest means of success was the employment of the aerial screw—" the sainted screw," as an illustrious mathematician called it, which was known to be capable of carrying up a mouse, and must, therefore, *à fortiori*, be able to sustain an elephant.

The inanity of this argument was not apparent at the time ; and *Nadar* proceeded to form a syndicate to promote " aviation" after the methods of opera bouffé. A journal was founded—the first *Aéronaute*—43 paying subscribers were obtained, and 100,000 copies of the first issue printed. This journal expired after its fifth issue. Then a monster balloon was built—the *Géant*—out of the exhibition of which it was expected to realize sufficient profits to build a screw machine which should put an end to ballooning forever. But the *Géant* met with all sorts of mishaps ; it gave no profits, and entailed losses instead, which nearly ruined *Nadar ;* and such experiments as were tried with aerial screws (outside of the little toys which were exhibited at the various meetings) demonstrated that the utmost weight which the exertion of one horse power could sustain, with a screw acting upon the air, was some 33 lbs., or, in other words, that if the apparatus were to weigh one ton, it would need 67 horse power continuously exerted to keep it afloat.

This is now clear enough to us. Assuming that in consequence of the rotation at high speed a smaller surface is required to sustain a given weight with a screw than with reciprocating wings or fixed aeroplanes, yet the motor for the screw would probably weigh about one-third of the whole weight of the apparatus (instead of one-quarter, as in the case of birds, and probably one-sixth in the case of aeroplanes), and so the utmost weight available for the motor of the screw and its supplies would be $\frac{1}{3}$ of 33 lbs., or 11 lbs. to the horse power, while in 1863 there was no primary motor known then approximating such phenomenal lightness.

Now that Mr. *Maxim* has announced that he has built a steam engine, and its generator of 950 lbs. aggregate weight developing 120 actual horse power, or at the rate of 8 lbs. to the horse power, it is doubtless within his power to go up into the air with an aerial screw, and to perform therein various evolutions ; but his trips would probably be short, and the consequences might be unpleasant were the machinery to break down while he is aloft.

He has, accordingly, with great good judgment, begun by applying his steam engine to an aeroplane, although this will involve greater difficulties in starting and in landing, as well as a less immediate demonstration.

Almost the only memento which now remains of the movement in favor of the aerial screw inaugurated by *Nadar* is the model of the flying machine designed by the *Viscount de Ponton d' Amécourt,* and which is shown in fig. 28. The following description is translated from that of M. *Tissandier :*

"M. *de Ponton d'Amécourt* constructed, in 1865, an aerial screw machine worked by steam, which was expected to rise with both its motor and its steam generator. This beautiful little model, which was exhibited at the London Aeronautical Exposition in 1868, is exquisitely finished. The boiler and frames are of aluminium, and

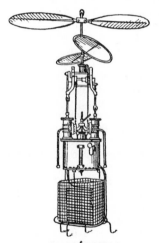

FIG. 28.—D'AMÉCOURT—1863.

the steam cylinders are of bronze. The reciprocating movement of the pistons is transmitted by gearing to a double pair of superposed screws of 41 sq. in. surface, one rotating in a different direction from the other. The apparatus, which is now in the collection of the French *Society for Aerial Navigation,* weighs, without water or fuel, 6.1 lbs. The boiler is 3¼ in. high and 4 in. in diameter ; the total height is 24½ in. Unfortunately the boiler cannot be worked at sufficient pressure ; when the machine is put into motion it possesses a certain ascensional force ; it loses weight, but it does not rise."

The illustrated papers also published about 1865 views

of a great steam flying machine, attributed to M. *de la Landelle.* These showed a hull flanked with aeroplanes, and surmounted with two masts, each carrying four sets of screws, and also a partly folded umbrella, presumably to open into a parachute. It is to be found reproduced in most works upon aerial navigation, and in encyclopædia articles, and is not given here, because it possesses no merit whatever, being probably a newspaper fancy, like the flying ship attributed to Mr. *Edison,* which went the rounds of the press some years ago, and which is also reproduced in M. *Dieuaide's* chart.

M. *de la Landelle* was a persevering man, as well as one of considerable scientific acquirements. He continued making experiments upon screws of various shapes long after MM. *Nadar* and *Ponton d'Amécourt* had given them up in disgust ; and he encouraged M. *Pénaud,* then a ris-

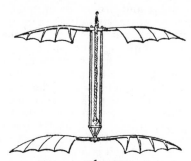

FIG. 29.—PÉNAUD—1870.

ing young man, to take up the study of Aviation. The latter produced in 1870 the little apparatus shown in fig. 29, which has remained the best of its kind.

Pénaud's flying screw, which is called by the French a " Hélicoptère," consists of two superposed screws rotating in opposite directions, and actuated by the force of twisted rubber strings. The principle is the same as the apparatus of *Launoy & Bienvenu* and of Sir *George Cayley,* but the twisted rubber is far more effective than the bow, whether the latter be of whalebone or steel, and this change in the motor constituted the chief merit of *Pénaud's* modification. He first experimented with rubber in tension, but found that the increased weight of the frame (to resist the strains) more than compensated for the weaker effects of torsion, and that the latter application enabled him to construct models which were simple, cheap, efficacious, and not easily broken. These models, when built in varying proportions, would either rise like a dart to a height of some 50 ft., and then fall down, or sail obliquely in great circles, or, after rising some 20 or 25 ft., hover in the same spot for 15 or 20 seconds, and sometimes as many as 26 seconds, which was a much longer flight than had ever before been obtained with screws.

For lack of a sufficiently light primary motor, *Pénaud's* further experiments in this direction brought forth no practical results, and his apparatus has remained a toy, which has been varied in many ways.

The most popular of such toys have been the various single spinning screws, either of cardboard or metal, which are attached to a spindle around which a string is wound, and which are set in rapid rotation by briskly pulling and unwinding the cord. These are of many shapes, with two, three, or more arms, and of various angles of pitch. Those with heavy rims are most effective, sometimes rising as much as 200 ft. into the air ; but they have led to accidents and proved dangerous. In such devices the source of power is not taken up into the air, as in *Pénaud's* apparatus, but it is stored in the momentum of the screw and encircling ring (if any) by the original muscular effort. Mr. *Wenham* measured the force expended in unwinding the coiled string by attaching thereto a small spring steelyard, and noting the time of ascent of a flying screw of tin plate with three equidistant vanes. He computed that to maintain the flight of the instrument, weighing 396 grains, a constant force is required of near 60 foot-pounds per minute, or in the ratio of about 3 horse power for every 100 pounds.

Many modifications have also been made of the double screw arrangement, which takes up its secondary motor in the shape of twisted rubber. These have been produced by many people ; but the cleverest are probably those of M. *Dandrieux*, who, in November, 1879, presented before the French Aeronautical Society * no less than 10 different types, the best known of which is that of the butterfly, which is still to be found in the toy shops, and which comes to us both from Paris and from Japan. M. *Dandrieux* modified the shape and proportions of the screw, and effected a material improvement in its efficiency.

Flying screws driven by clock springs have been frequently made. Such an arrangement was constructed by Sir *George Cayley*, "the flying baronet," at the beginning of the century, and is described in his paper on "Aerial Navigation," in Vol. XV of *Nicholson's Journal.* Sometimes the attempt has been made to substitute man power. Of such was the experiment of Mr. *Mayer*, a

---

* *L'Aéronaute*, January, 1880.

stair-builder, who, about 1828, constructed an aerial screw proportioned to sustain 126 lbs., and rotated it with his own muscular power. In giving an account of the result, forty years later, he said, naively : *

"The result was very flattering, though not perfectly successful. My pecuniary resources were exhausted, and other work in my own business being then offered to me, ascending by wings (screws) was abandoned until a more convenient season, and the more certain and substantial method of making stairs, and ascending them step by step, was substituted in its place."

Realizing the utter insufficiency of man power, or of any known primary motor, some inventors have designed flying screws to be worked by new-fangled motors. Of these was the apparatus of *Pomès & de la Pauze*, proposed in 1871, and shown in fig. 30. The sustaining screw

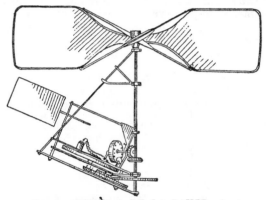

FIG. 30.—POMÈS & DE LA PAUZE—1871.

was inclined so as to obtain an oblique ascent, and appears to have been adjustable. The steering was to be done by a rudder, and the whole was to be worked by a gunpowder motor. The first requisite, therefore, was to perfect the gunpowder engine. It is not known how much was accomplished toward this ; but the flying apparatus was never built.

The next year (1872) M. *Renoir*, a member of the French Society, proposed an apparatus consisting of two aerial screws placed side by side in the same horizontal plane, but with shafts capable of being moved out of the vertical, in order to secure movement in both directions. They were to be driven by steam, and to rotate in opposite directions ; and M. *Renoir* computed that the axis

* Third Annual Report Aeronautical Society of Great Britain.

of rotation would have to be inclined 11° in order to obtain a horizontal course. Also, that to produce satisfactory forward speed, the additional power required would be but 10 per cent. of that required for sustaining the weight. Aside from the main question of the motor, which was left in abeyance, the important thing to ascertain was the best form of sustaining screw, in order to get the utmost support with the least expenditure of power ; so the succeeding year, M. *Renoir*, having studied the results obtained by M. *Pillet* with a concave screw * in a series of experiments beginning in 1848, tried some experiments of his own with a screw provided with a return flange or turned edge, to prevent the centrifugal escape of the air, of which he gave an account in the *Aéronaute* for April, 1873.

He drove his screw by man power, and claimed that the results showed that a force of one horse power could sustain, by means of his screw, a weight of 165 lbs. ; but Mr. *Bennet*, in giving an account of these experiments to the Aeronautical Society of Great Britain, in 1874, gave a somewhat different account, and said :

Two years ago M. *Renoir*, a member of the French Society, experimented with a screw 15 ft. in diameter, with which, by the action of his feet, he was able to lift a weight of 26 lbs. The screw was two bladed, with an increasing pitch, the angle of inclination being 3° at the front edge of the blade and increasing to 30° at the back edge. The two blades cover the entire area of the screw, and have a deep rim suspended from them to prevent the air being driven from the circumference by centrifugal force. M. *Renoir* estimated the power he developed was about one fifth of a horse power ; but this was considered by the members of the French Society present at the experiment to be considerably below the real power exerted. As the screw was driven by the feet, after the manner of a velocipede, the body being in a good position for exerting its maximum effort, the power developed was undoubtedly nearly one horse power. A man running up a pair of stairs is able for a few seconds to exert two horse power, and mounting a ladder placed vertically, by the help of his hands, an ordinary man can do the work of 1¼ horse power. These facts have been determined by experiment.

While on the subject of the form of screws, it may be well to call the attention of those who may desire to study the subject further to an article upon "Propulsors," by M. *Crocé Spinelli* (the same gentleman who lost his life in the scientific balloon ascension of the *Zenith*), which will be found in the *Aéronaute* for April, 1870, and to another

---

* *Aéronaute*, March, 1870.

by the same author on "A Screw with Variable Pitch" in
the *Aéronaute* for November, 1871.   Also to the remarks
on Screws by Mr. *Wenham* in the first and second reports
of the Aeronautical Society of Great Britain, and to those
of Mr. *Thomas Moy,* in the fourth report of the same society.   He evidently knew what he was talking about.

In 1872 Mr. *Wenham* proposed a method for varying
the pitch of the screw, which may be found in the report
of the British Aeronautical Society of that year.   The
blades were to be made of some fabric, one edge being
attached to a cross arm, which was made fast to the shaft
of the screw.   The other edge of the fabric was fastened
to another cross arm, so arranged as to be placed in any
position on the shaft, and firmly fixed in such position.   A
coiled spring was to keep the two cross arms apart, and
thus maintain the fabric tightly stretched.   If the adjustable arm be placed precisely in line with the fixed arm,
then the blade is parallel with the shaft, and by moving

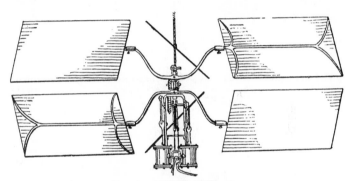

FIG. 31.—DIEUAIDE—1877.

the adjustable arm to one side more or less, the pitch can
be made anything desired.

The next experiments on screws were tried in 1877, by
M. *Dieuaide,* formerly Secretary of the French Aeronautical Society, and the well-known Engineer and Patent At-
torney, whose clever *chart* has furnished (by permission)
almost all the illustrations contained in these articles.
His apparatus is shown in fig. 31.   It consisted of two
pairs of square vanes set at various angles to the line of
motion, so as to vary the pitch, and rotated in contrary
directions by gearing.   The power was furnished by a
double cylinder steam-engine connected with the boiler by
a flexible hose, and the lifting power of the screws could
be accurately weighed by simply putting the apparatus on
a scale.

The results of the experiments seemed to show "that this double screw could not, in consequence of the losses of power due to the gearing, exert a lifting force greater than that of 26.4 lbs. per horse power." This agrees closely with the results of the experiments of *Giffard* with a single screw ; he having found that 6 horse power would lift with a screw 165 lbs. at the rate of 3.28 ft. per second, or say 27.5 lbs. per horse power, from which he deduced the conclusion that the aerial screw gave out but 18 per cent. of the power exerted to drive it.

The next apparatus to be noticed was not experimented with, so far as the writer has ascertained, but was a proposal of great oddity and originality patented in 1877 by M. *Mélikoff*, Engineer and graduate of the school of the "Ponts-et-Chaussées." It is shown in fig. 32, and consisted in a sort of screw parachute composed of "two hyperbolic paraboloids united by their concavities into a

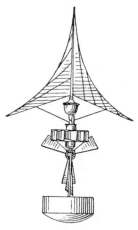

Fig. 32.—MÉLIKOFF—1877.

sort of cone or pyramid with a rectangular base in projection." This was to be furnished with a series of zones, shown in section in the figure, to act upon the air ; and this arrangement, the one resembling a spear-head in the figure, was expected to screw itself up into the air and to act as a parachute in coming down. It was to be rotated by a gas turbine, consisting of eight curved chambers, into each of which charges of the vapor of ether mixed with air were to be successively exploded by an electric spark, and the charges allowed to expand in doing work. The surfaces were to be kept cool by melting ice and by heating the resulting water. This ice and the supply of ether were to be carried in the recipient shown just below the parachute, the turbine being shown lower down ; this motor was expected to work also an ordinary screw with three arms, geared on a short axle, from which screw horizontal propulsion was expected. Below all is shown the car for the operator.

M. *Mélikoff* designed his apparatus to carry up one man, and estimated its total weight at 374 lbs. Of this the apparatus proper was to absorb 108 lbs., the gas turbine was to weigh 92 lbs., its supplies for one hour were to amount to 40 lbs., and the operator was to be of 134 lbs. weight. The rotating surface was to measure 87 sq. ft. in area, thus giving a proportion of 4.3 lbs. to the sq. ft., which seems entirely too small, although claimed to be calculated from the tables of air pressures given by *Thibault.* The turbine was to be of 4 horse power, being thus estimated to weigh 23 lbs. per horse power, and it was to consume per horse power per hour 3.3 lbs. of ether and 8.7 lbs. of ice for cooling the parts, thus showing a slight discrepancy from the aggregate of 40 lbs. of supplies estimated as required for one hour.

The apparatus as a whole is scarcely worth experimenting with, and has been chiefly described because of its oddity ; but the weight and power of the projected gas turbine seem to have been worked out with some care, and it might be worth while to take the subject up again, in order to ascertain whether it is practicable to construct a rotary gas motor weighing as little as 23 lbs. to the horse power.

The next experiment to be noticed was tried by M. *Castel*, a mechanical engineer, in 1878. He wanted to deter-

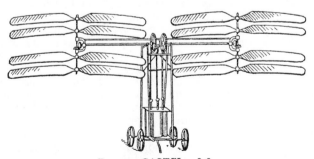

FIG. 33.—CASTEL—1878.

mine the amount of mechanical work required to sustain a motor in the air, and built the apparatus shown in fig. 33. It consisted of eight double screws rotated in opposite directions by a double-cylinder compressed-air engine, mounted upon wheels and fed with compressed air through a long, very light rubber hose. The weight of the whole apparatus was 49 lbs., of which 22 lbs. was in the screws and their machinery. The screws were 3 93 ft. in diameter, and weighed 1.32 lbs. each.

Experiments were repeatedly tried, but they came to an early ending by the apparatus rising upon the air, taking a sheer, and smashing itself against the wall of the room. M. *Castel* did not publish the results accomplished in the way of lifting a measured number of pounds per horse power developed ; but he stated that he "no longer had the confidence which he once possessed in screws as future instruments of aviation. Elastic surfaces with an alternating action to impart vibratory motion to the air now seem preferable to screws to solve the problem of aerial navigation with an apparatus heavier than the air." He estimated from an examination of the muscles of birds and of the amount of work which those muscles were able to give out, that the bird in full flight expended not more than 24 foot-pounds per minute for each pound of his weight, so that a bird, if he weighed 220 lbs., would only expend a maximum of 0.16 horse power.

Now, we have already seen that the average power of a man is 0.13 horse power, and that although he weighs less than 220 lbs., he cannot fly with wings by his muscular efforts, so that the estimate must be erroneous.

M. *Castel* proposed to build a petroleum motor to drive his proposed wing apparatus, but he probably found himself unable to keep within the necessary limits of weight.

A simpler apparatus than M. *Castel's* accomplished much better results, for in the same year (1878) Professor *Forlanini*, an Italian civil engineer, launched into the air the second steam apparatus which has flown with its contained supply of steam ; the first having been that of Mr. *Phillips*, already described. Fig. 34 shows the flying screw arrangement experimented with by M. *Forlanini*.

It is composed of two double-bladed screws, of which the lower one is rigidly fixed to the steam-engine, while the upper one rotates ; the result being that the lower screw furnishes a fulcrum upon the air, while the upper one furnishes the ascending power. The whole apparatus thus slowly rotates upon its own axis ; but this feature, which would be very objectionable in a really navigable apparatus, could be eliminated by rotating both screws in inverse directions.

The upper screw was worked by a double cylinder steam engine of ¼ horse power, supplied with steam from superheated water contained in a depending hollow globe after the manner of the well-known fireless locomotive, the initial pressure being some 120 to 160 lbs. per sq. in. It was the original design of M. *Forlanini* to send up his apparatus with a steam boiler attached, fired by 200 minute alcohol flames ; but this proved too heavy to be lifted by the machine, and he substituted the hollow globe, tested to an internal pressure of 225 lbs. per sq. in., which, being two-thirds filled with water, is simply laid upon a fire until the desired pressure is obtained ; when, on being withdrawn, the throttle-valve which admits steam to the cylinders is opened, and the apparatus rises.

It has been repeatedly tested, and its best performance seems to have been to rise to a height of 42 ft. and to remain 20 seconds in the air. M. *Forlanini* expressed the intention of following it up with an improved apparatus, of which he had the design, and with an engine of 2 horse power ; but it is stated that he has not had the leisure to carry out this intention.

The total weight of the original apparatus was 7.7 lbs., and the aggregate area of the screws was 21.5 sq. ft., thus giving a bearing surface of about 2.8 sq. ft. per pound. The weight of the steam-engine proper was 3.52 lbs. and that of the screws 1.32 lbs. The hollow globe, charged with water, weighed 2.20 lbs., and the steam-gauge and connections weighed 0.44 lbs. more, leaving 0.22 lbs. for other accessories. It will be noticed that the engine, the

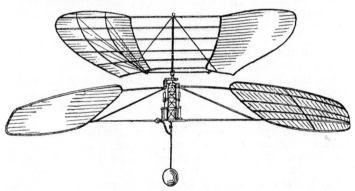

FIG. 34 —FORLANINI—1878.

boiler and the gauge weigh about 80 per cent. of the whole, which proportions could not be expected to obtain in a navigable apparatus ; but, on the other hand, a larger steam-engine and boiler would weigh less in proportion to its power than the minute one thus experimented with, in which steam was very wastefully used in consequence of the relatively very large proportion of radiating surfaces.

M. *Forlanini* designed a self-generating steam boiler,

which he expected to weigh but 13.2 lbs. per horse power ; but it is not known to have been constructed.

This, then, is the best that has hitherto been done with steam. A model screw machine weighing 7.7 lbs. has risen 42 ft. into the air and flown for 20 seconds, but without taking up a self-generating steam boiler. The power developed ranged from 7,800 to 10,850 foot-pounds per minute, and the total weight sustained was at the rate of 26.4 lbs. per horse power.

Some time about the year 1880 Mr. *Edison*—the great Edison—at the instance of Mr. James Gordon Bennett, made some preliminary experiments to promote aerial navigation. He began very judiciously by trying to ascertain what could be done with the aerial screw as a propeller. For this purpose he is reported to have placed an electric motor of 10 (?) horse power, connected with a vertical shaft surmounted with rotating vanes upon a platform scale, and to have connected it by a wire with a source of electric power—the object being to ascertain how much the whole could be lightened by the action of the vanes upon the air.

He rigged upon the shaft first one kind of propeller, and then another, until he had tried all that he could think of ; the best being a two-winged fan with long arms.

He is reported as saying that the best results obtained were to lighten the apparatus some four or five pounds of its total weight of 160 lbs., but the amount of power developed is not stated. This must have been quite small, and Mr. *Edison* must have been unfortunate in his selection of the screws to be tried, for we have seen, by the experiments of others, that a motor of 10 (if it was really this) horse power ought to lift 260 lbs. It is no wonder that he is reported as saying that "the thing never will be practicable until an engine of 50 horse power can be devised to weigh about 40 lbs."

It is understood that somewhat similar experiments were tried by Mr. *Dudgeon*, the celebrated maker of hydraulic jacks. He tested the lifting effect of various forms of screws when rotated by steam power, and, like Mr. *Edison*, he stopped in disgust when he found how small was the lift in proportion to the power expended.

There may have been other experiments with lifting aerial screws in the United States, but they have not come to the knowledge of the writer. In point of fact, such aerial devices do not seem to have received much attention from inventors, and there have been but few patented proposals therefore in the United States.

In 1876 a patent was taken by Mr. *Ward*, of San Francisco, for an aerial vessel in which the supporting and the propelling power was to be furnished by a series of fan blowers. The fans furnishing the support were placed on horizontal shafts and the exhaust opened downward, so that the reaction would act against the force of gravity, while the fans which produced the horizontal motion were also arranged on horizontal shafts at the rear, the air being conducted to them through a duct from the front, and exhaust being to the rear, so that the reaction would force the vessel forward.

In 1877 Mr. *Ward* took out further patents, in which the apparatus was somewhat modified, but the general principles remained the same. It is believed that he tried some experiments ; but no record of them has been met with by the writer, and a letter to the inventor has remained unanswered.

The same idea, but in a modified form, has quite lately (1892) been patented by Mr. *Walker*, of Texas ; and perhaps experiments will be tried to test the lifting effect of air blasts under favorable circumstances ; but as the efficiency of a screw, when used as a fan, is stated at only 35 per cent., while its efficiency as a propeller is stated at 70 per cent., it seems a question whether air blasts can be advantageously used in aerial navigation.

It may be pointed out here that there is a considerable difference between the fan blower and the screw propeller —a difference which should be more thoroughly understood by inventors. The most efficient fan blower is a machine which will produce the strongest current of air with any given expenditure of power. The best screw propeller is the machine which will produce the least current. If a screw propeller could be so arranged that it would not put the air in motion at all, then there would be no " slip," and the machine would be as efficient as a locomotive running on a dry rail, in which case all the power is expended upon the vehicle. In the case of a fan blower, or in the case of a steamboat moored to the wharf, and with its engines in operation, all of the power is expended in moving the fluid. It is all wasted in slip. In the case of the steamboat advancing through the surrounding fluid, or of the aerial machine, if it ever gets under way, a part of the power is expended in putting the craft in motion and another part in putting the fluid in motion, and the latter power is inefficient ; it is the " slip." The best screw, therefore, is the one which shall expend the greatest part of the applied force upon the craft and the least upon the fluid. It is the screw which will create as little movement as possible in the fluid in which it operates.

In 1879 Mr. *Quinby* patented a device consisting of two sets of screw-like sails, one set to raise the machine and

the other to propel it.   The drawing shows a light framework with two screws, each with two blades of fabric,
one set on a vertical mast, and the other upon an inclined
mast.   The screws were to be driven through rope gearing by some source of power.

In the same year Mr. *Greenough* also patented an apparatus, which should better, perhaps, be noticed under the
head of aeroplanes, but which differed from this type by
having lifting screws imbedded in the surface of the aeroplane, in order to obtain a lifting action upon first getting
under way, after which, by sailing at an angle, both sustaining and propelling effect could be obtained from the
screws, with, however, the possible addition of a vertical
screw to give increased forward motion.   This inventor is
understood to have tried some preliminary experiments of
details, and as a result thereof to be awaiting the development of a light motor before undertaking to realize his
conception upon a navigable scale.

In 1885 Mr. *Foster* patented an air ship consisting of
two screws, four-bladed, side by side, on separate vertical shafts, which latter can be thrown at an angle by reason of a flexible portion connecting with the main driving
shaft, so that the thrust may both lift and propel the apparatus.   The main shaft was to be driven by the feet of
an operator sitting below and half way between the two
screws.   These screws are apparently some 8 ft. in diameter, and the man power relied upon is evidently inadequate, so that it is quite safe to say that if the apparatus
was ever tried it did not succeed in rising.

In 1886 and 1887 some experiments were tried at the
Royal Dock Yards in Copenhagen, for the purpose of determining the relative efficiencies of screws operating
in water, and those which should operate in the air.
The experiments were in connection with marine, and not
with aerial navigation ; but it was found that not only
would the aerial propeller develop as great a thrust as the
water propeller, in proportion to the energy consumed,
but that under certain conditions it would do slightly more,
and greater thrusts per horse power were attained than in
any previous experiments.

These very important results, for which most of our readers will be unprepared, warrant noticing the experiments
at some length.   They were described in a paper by *H. C.
Vogt*, read before the British Association in 1888, and
the first seems to have consisted in the careful measurement by Mr. *Freninges*, of Copenhagen, of the thrust and
work done by a largish flying screw, two-bladed, 1 ft. in
diameter and 1 ft. pitch, weighing 0.35 lbs.   With 70
revolutions per second, it will rise 200 ft. into the air, and

Mr. *Freninges* determined the efficacy or work done to be 63 per cent. of the kinetic energy imparted by the arm of the operator. At 52 revolutions per second, requiring the expenditure of 100 foot-pounds, the thrust of the screw against a stop was 6 lbs., and its efficiency therefore was $\frac{6 \times 550}{100} = 33$ lbs. per horse power, which agrees well with the measurements of Mr. *Wenham* and others.

The first dock-yard experiments were undertaken by Messrs. *Dahlstrom & Lohman*, and consisted in fitting out a launch 20 ft. long and 5½ ft. beam with an aerial screw propeller of canvas 8½ ft. in diameter, having 24 sq. ft. of area distributed over two ordinary canvas sails, the pitch of which could be varied. The engine was 1½ horse power. Measured by a spring balance, the thrust of the air propeller was, in calm weather, 36.7 lbs. per indicated horse power, or the same as that of a water-screw turned by the same power. In windy weather this thrust was augmented through 75 per cent. of the directions in which the wind could blow, thus illustrating the fact that if a current of air be blowing across the blades the efficiency of a propeller will be increased, because many more particles of air will be acted upon in the same space of time than in a calm. This fact promises important consequences for an aerial screw in propelling, should a true flying machine ever be compassed, for then the advancing screw would constantly have fresh particles of air to work upon, and there would be a reduction in the slip which necessarily must occur when its thrust is measured in a fixed position.

The next experiment was tried with the Government Dock Yard launch, which was 31 ft. long and 8 ft. beam. Its ordinary water screw was removed, and an air propeller of canvas was substituted, which was 20 ft. in diameter and had a total area of 250 sq. ft. This area was found much too large, but by reducing it to about 150 sq. ft. an average speed of nearly 7 knots was attained by the launch, whose speed with the ordinary water screw and the same power (11.3 indicated horse power) was a maximum of 7.3 knots per hour. There was, however, a slip of the driving rope which was estimated as wasting about 2 horse power, and the director estimated that the speed with the air propeller would have been 7.5 knots per hour if the gear had worked properly. As on previous trials, 75 per cent. of the winds increased the thrust of the propeller.

The apparatus for the next experiment, which was tried in 1887, was made by Messrs. *Dahlstrom & Lohman*, engineers, of Copenhagen. An air propeller with three vanes of thin sheet steel, and an area of about 5 sq. ft., was fitted to a boat 16 ft. long and 4½ ft. beam, and rotated by man power. It is stated to have produced a thrust of 10 lbs., with an effort of about 100 foot-pounds, or at the astonishing rate of 55 lbs. per horse power ; but it must have been assisted by wind blowing athwart the blades, for Mr. *H. C. Vogt*, in a letter published in London *Engineering* for December 4, 1891, says, in discussing Aerodynamics, that "with 1 indicated horse power it is not possible to obtain a thrust of over 40 lbs. to 45 lbs. with an air propeller— say 50 lbs. to 60 lbs. per brake horse power on the shaft —just the same in whatever manner area, pitch, and revolutions are varied."

On the basis of these Copenhagen experiments Mr. *John P. Holland*, in a very interesting letter, published in the New York *Herald* in November, 1890, claims that it is even now possible to navigate the air upon the screw principle, by simply combining things already tried and proved by various experimenters ; and he gives the elements of a proposed steam apparatus, weighing some 7,000 lbs., and capable of carrying two men, with supplies of fuel, etc., sufficient to sail from 8.44 to 23.6 hours. Details of the design and method of operation are withheld until a patent can be secured. As has already been said in referring to Mr. *Maxim*, it is probable that such a machine can be made to rise upon the air ; but special appliances will be required to secure safety in case the machinery breaks down while under way, and in effecting a landing.

A somewhat similar proposal is made in a pamphlet published in 1891 by Mr. *James Means*, of Boston, but he gives only a scanty glimpse of the arrangement by which he thinks the problem could be solved. He proposes one screw on a vertical shaft, sustaining a car, with a pair of widely extended vertical planes, to prevent rotation of the apparatus, and concludes by saying : "If you want to bore through the air, the best way is to set up your borer and bore."

Our knowledge of the action of aerial screws is almost wholly experimental ; and it would seem, in the present chaotic state of theory as applied to the screw, as if this remark of Mr. *Means* was almost as comprehensive and reliable as anything on the subject of aerial screws which has been published up to the present time. The writer feels quite certain that it contains in a condensed form as much reliable detailed solid information as several mathematical articles of considerable complexity which he has consulted, and it will be seen, by closely analyzing Mr. *Means's* suggestion, that after its entire adoption in the spirit in which it is made, there would be little left to be desired in the development of aerial screws.

Among the inventors who have most deeply and most intelligently studied the action of screws must be mentioned M. *G. Trouvé*, of Paris, whose artificial flapping bird has already been noticed under the head of " Wings." He has proceeded almost wholly in the experimental way, and he has accomplished some very remarkable results. He began his experiments with marine screws applied to electric launches about 1881, and soon developed an electric motor weighing but 33 lbs. per horse power (primary battery not included), which rotated an improved marine screw some 2,400 turns per minute.*

In 1886 he exhibited to the French Academy of Sciences a new method of constructing geometrically accurate screws by a process so simple that any workman can carry it out, and that the cost is very much reduced. He has also experimented, ever since 1867, with aerial screws, and has reached the conclusion that for the latter

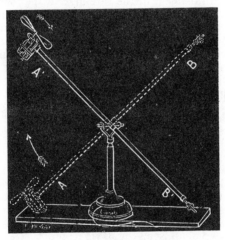

FIG 35.—TROUVÉ—1886.

the best results are obtained when the pitch is equal to the diameter, or a little less,† contrary to marine practice, where pitch is generally 1.3 times the diameter.

In 1887, at the Scientific Congress at Toulouse, and in 1888, before the French Société de Physique; M. *Trouvé* exhibited the electric motor and aerial screw represented in fig. 35. The motor is the lightest ever built, weighing

---

* Histoire d'un Inventeur—Barral. Page 416.   † *Ibid.* Page 442.

but 3.17 oz., and developing 868 foot-pounds per minute, or at the astonishing rate of 1 horse power for each 7.42 lbs. weight. It is wholly of aluminum, except the magnetic circuit, which is necessarily of very soft iron ; and the armature is directly connected with a very light aerial screw, geometrically perfect, which was constructed by the process communicated to the French Academy of Sciences.

This apparatus, upon being placed in one pan of a pair of scales, and connected with a source of electricity of 40 Watts constant delivery, lightened itself of its entire weight by action upon the air. To make the experiment more striking, M. *Trouvé* then arranged it at the extremity of a balanced beam, as shown in the figure, connecting it with the electric supply through the standard, the knife edges and the beam. Then upon turning on the current, the screw began to revolve, and the balanced beam rose from the position *A B* into the position *A' B'*, with the expenditure of 868 foot-pounds per minute, which M. *Trouvé* says is capable of raising it at the rate of 72 ft. per second.

Inasmuch as he estimates that this minute motor has only an efficiency of 20 per cent., and that a similar one of 50 to 100 horse power would possess an efficiency of 80 to 92 per cent., it would seem that M. *Trouvé* now has it in his power to go up into the air with a pair of aerial screws, rotating in contrary directions in order to insure stability, moved by his wonderfully light motor, to float, to hover, and to move about at pleasure so long as he remains within the limits of length, of strength and of weight of a connecting wire to convey the electric force from a dynamo and steam engine, which remain on the ground, to the electric motor and aerial screw in the air.

This, as he points out, would be of practical use on the battle-field or in a besieged city, to observe the enemy ; and it is not impossible that he will exhibit such an apparatus at some International Exposition ; but he believes that he has now designed a still better solution of the problem ; and we shall see, when we come to treat of aeroplanes, that he made the plans for an apparatus of that kind which seems to him to solve, both in arrangement and motive power, the all-important question of the navigation of the air.

For several years past series of experiments upon aerial screws, both for sustaining and for propelling, have been carried on by *Commandant Renard*, at the French Aeronautical War Establishment at Chalais. He published a preliminary paper in the *Revue de L'Aéronautique*, in 1889, in which he gave a description of the machine used in testing, and of the results of the experiments with the screw of the war balloon *La France*, which is two-bladed, nearly 23 ft. in diameter, with an average pitch of 27.5 ft. and a surface of about 42 sq. ft.

He found that the efficacy of this screw, or its thrust in pounds divided by the foot-pounds exerted, varied from 48.4 lbs. per horse power at 17 turns per minute, down to 16.94 lbs. per horse power, with 48 turns per minute ; and he calls attention to the fact that inasmuch as the thrust increases as the square of the velocity, while the power required grows as the cube, the proper method of comparing the efficiencies of various forms of screws is to compare the quotients obtained by dividing the cubes of the thrusts by the squares of the powers.

Commandant Renard seems to have so proceeded in comparing his experiments ; and in a paper read by him before the French Society of Physics, in 1889, he stated that of seven forms of screws tried up to that time, one was much better than the others ; and he added from theoretical considerations : " There must be a screw for which $\dfrac{\text{Thrust}^3}{\text{Power}^2}$ = constant, is a maximum. This is confirmed by experiment ; and it shows, moreover, that this maximum when plotted resembles a sharp peak, each side of which forms a *veritable precipice*. In other words, there is a screw very much better than others, and its form cannot be much departed from without producing very bad aerial screws."

None of the forms of screws experimented upon are published, save that of *La France*, and that this is not the best may be inferred from the fact that Mr. *Maxim*, who tested about fifty different forms of screws in his recent experiments, says : " The screw which gave the worst results was made exactly like those employed in the experiments of the French Government."

Mr. *Maxim* has published a popular account, all too brief, of his experiments, in the *Century Magazine* for October, 1891, but for obvious reasons does not go into scientific details. He has expressed the intention of eventually doing so, and this is sure to prove a very great addition to our present scanty knowledge, for his experiments on aerial screws have been more systematic and comprehensive than any heretofore tried.

From the foregoing it will be seen that comparatively few experiments have been made to compass artificial flight by means of sustaining aerial screws, and that much, very much remains to be learned concerning the best form to be given to them, the proper area, velocity and pitch, as well as the power required, either for sustaining or for propelling a given weight in the air with a screw.

Indeed, even for marine screws, our knowledge may be said to be wholly empirical—that is to say, based on experiment ; and there is no mathematical theory of them which has found general acceptance, or which connects their action with that of plane surfaces, so as to agree with the observed facts. Some calculations made by the present writer seem to indicate that it may be less difficult to do so, in the case of aerial screws ; but it must be acknowledged that we really know but little about them, and that the most that we can say at present is that while a flying machine in which the sustaining power is to be obtained from rotating screws is likely to require less surface than an aeroplane to sustain the same weight, perhaps in the proportion of about one-third, yet it is likely to require more power than the aeroplane to obtain the same speed of translation, and also to involve greater risks of accidents in case of failure of any part of the machinery.

It would seem to the writer as if the true function of aerial screws was to propel, leaving the sustaining power to be obtained in some other way, and we will therefore pass to the consideration of AEROPLANES.

AEROPLANES.

AEROPLANES—*i.e.*, thin fixed surfaces, slightly inclined to the line of motion, and deriving their support from the upward reaction of the air pressure due to the speed, the latter being obtained by some separate propelling device, have been among the last aerial contrivances to be experimented upon in modern times.

The idea of obtaining sustaining power from the air with a fixed, instead of a vibrating or a rotating surface is not obvious, and it was not till 1842 that an aeroplane, as we now understand the term, consisting of planes to sustain the weight, and of a screw to propel, was first proposed and experimented with. All aviators must have occasionally seen and marveled at the performances of the soaring varieties of birds, sailing in every direction at will upon rigidly extended wings (a performance concerning which more will be said in the progress of this discussion), but the flapping birds are so much more numerous and easily observed, their action is so much easier of comprehension, that they have been the favorite model.

We shall see, however, in reviewing old traditions with perhaps a new understanding, that such faint approximations to success, as have hitherto been attained with artificial flying machines, were probably accomplished with fixed surfaces, either by gliding downward by the force of gravity, or in soaring upon the wind like a bird.

Although aeroplanes have been among the last devices to be experimented upon, they are now the favorite apparatus from which success is hoped for, and later designs have chiefly been in this direction. The very important labors of Professor Langley have shown that with the exertion of 1 horse power as many as 209 lbs. can be sustained in the air by an aeroplane, while Mr. *Maxim* states, as the result of his many experiments, that at least 133 lbs. can be sustained per horse power. If we compare this with the results of vibrating wings, which may be assumed as supporting 77 lbs. per horse power,* or with screws, which have been shown as affording generally 33 to 40 lbs. per horse power—chiefly, as the writer believes, because of the greater angle of incidence or of pitch which is required with alternating or rotating surfaces—we see that fixed surfaces possess a marked advantage over movable surfaces. The latter, it is true, can probably be made somewhat smaller and hence lighter, but this advantage is likely to be more than counterbalanced by the greater power required to work them.

Still the fact remains that almost all experiments with aeroplanes have hitherto been flat failures. This is believed by the writer to result from the difficulty of maintaining the equilibrium of that form of apparatus, both sideways and fore and aft. There seems to be no very great difficulty in obtaining proper balance and equilibrium with flapping wings or with screw apparatus, the motion of the parts apparently compensating to some extent for the tendency to tilt over, but fixed aeroplanes seem much more unstable—a single flat plane, for instance, of uniform weight throughout, possessing no sort of stability whatever when in forward rectilinear motion.

Perhaps the reader will best understand this, and at the same time obtain a glimpse of some of the laws of aerial equilibrium under forward motion, by experimenting with an aeroplane of his own.

Let him cut out a strip or parallelogram of stiff wrapping paper—say 15 in. long and 3 in. wide. Its sides should be straight and parallel, and the surface a true flat plane, free from folds or wrinkles. In this condition it may fall flatways a short distance with tolerable steadiness ; but if allowed to fall edgeways, or projected forward without whirling, it will at once rotate upon its long axis, tumble over and over, and be seen to have very unstable equilibrium. This can be remedied with very

---

* Assuming a bird in horizontal flight to develop 425 ft. lbs. per minute for each pound of his weight, the sustaining reaction will be $\dfrac{33.000}{425} = 77$ lbs. per horse power.

slight changes ; for next paste upon one of the long edges
of the plane a strip of pasteboard ½ of an inch wide and
the same length as the paper plane.   Part of an ordinary
pasteboard box will do very well ; but the important point
to be observed is that the apparatus shall balance on its
middle line, and at a point 28 to 30 per cent. back from its
front edge, or say, ⅞ of an inch.   When the paste is dry,
make a very slight fold in the paper strip near the edge
opposite to the pasteboard strip and parallel to it.   Let
this fold or crease be about ⅝ of an inch from the rear
edge, and form an angle of about 10° with the plane of the
paper.   Next bend the aeroplane parallel with the short sides
and exactly in the center of the long sides, so that the two
halves shall also stand at a diedral angle of about 10° with
each other, like a very obtuse letter V, care being taken
that middle fold and the back fold shall be on the same
side of the plane.   It will then be noticed that the attitude
of the apparatus somewhat resembles that of a soaring
buzzard in the air minus its tail.

If, now, this aeroplane be allowed to drop edgeways,
with the weighted side downward, from a height of 7 ft.
or more, its behavior will be entirely different from that
in its former condition.   Instead of tumbling over and
over, it will sail downward and forward upon a curve until
the increasing pressure balances the weight, and then
glide on a straight path to the floor, some 15 or 20 ft. from
the operator.   It may not glide on a straight line upon the
first trial, but in that event very slight changes in the
angles of the back crease and of the middle fold, and a
smoothing out of the plane will be sure to produce the desired forward flight and steady glide.

Still better results can be obtained by pasting a strip of
tin ¼ in. wide in a fold in the forward edge of a plane 4
in. wide.   For this purpose it will be well to have the
strip of tin 15 in. long, and to cut the paper plane 20 in.
long by 5 in. wide, so that 1 in. of the latter can be folded
quite over and pasted down over the tin.   The latter
should be accurately spaced 2½ in. from each end, so that
the apparatus shall balance exactly on the middle line
lengthways, and 1.2 in. from the weighted edge, cross-
ways.   The corners beyond the tin may be rounded off if
desired, provided care be taken not to disturb the balance.
Then by bending the apparatus very slightly in the center
of its length, and turning up the rear edge about 10° in
the same direction, an aeroplane will be produced which
will sail steadily forward in still air, sweep to the right, if
the right-hand back corner be slightly curved up, or go to
the left when the left-hand corner is similarly treated.

The principle on which this aeroplane sails is the same

as that upon which the bird glides downward on outstretched wings. The preponderance of the weight in
front determines the angle of incidence, and brings the
center of gravity to coincide with the center of pressure,
the latter varying approximately as per Joëssel law, already
given, which, however, it must be remembered, probably
only applies to square planes ; the horizontal component
of the pressure (inasmuch as the plane is inclined forwaid)
acts in the direction of the flight, and furnishes the motive
power while the back fold supplies automatically the longitudinal stability by counteracting such tendency as the
aeroplane may have to tilt fore and aft ; and the diedral
angle in the middle gives lateral stability, by reacting
against the air on that side toward which the apparatus
may begin to tip.

These compensations are effective in still air, but it may
be doubted whether they are sufficient in the open air.
When a bird soars in a gusty wind (and almost all winds
are gusty and irregular in velocity near the surface of the
ground), the automatic effects obtained by the diedral
angle of the wings and the upward angle of the tail do
not seem to act quickly enough. The bird will be seen,
by observation at close range, to be almost constantly balancing himself by slight, almost unconscious movements.
He advances the tips of his wings or thrusts them back ;
he flexes one or the other, and quite often he advances or
draws back his head, or uses his legs as a pendule from
the knee joint, in order to maintain his equilibrium. All
birds are acrobats, but the soaring kind, if closely observed in a gusty wind, will be seen to perform feats of
balancing more delicate and wonderful than those of any
human equilibrist.

We shall hereafter see that even if the aeroplanes experimented with had been provided with adequate motors,
as they were not, this difficulty in maintaining a proper
equilibrium with fixed surfaces is probably sufficient to
account for most of the failures of experiments upon a
practical scale with that form of apparatus, and for their
abandonment by their designers, a brief trial having probably satisfied them that aside from the question of a motive
power, which they were confessedly unable to solve, they
were not yet masters of such reasonable stability and com
mand over their apparatus, as to warrant them in proceeding further. Now that the all-important question of a light
motor seems to be in a fair way of being solved through
the achievements of Mr. *Maxim*, M. *Trouvé*, and others
who are known to be laboring in the same direction, the
question of the equilibrium of aeroplanes increases in
relative importance, and warrants making this somewhat

prominent in criticising past experiments and proposals. For this and other reasons, we shall pass in review a number of mere designs as well as forms of apparatus which was actually subjected to the test of experiment, and endeavor to inquire into the causes of failure.

Failures, it has been said, are almost as instructive as successes, as tending to remove, if we can understand the cause, at least one of the difficulties in the way, and the reader will probably agree that there has been hitherto no lack of failures in aerial experiments.

There probably have been in all ages of the world men, whose imaginations were fired by the sight of the soaring birds, and some who tried to imitate them. In early times mechanical and mathematical knowledge was too crude to render such experiments numerous, and before the invention and diffusion of printing, even the records of such failures would generally perish ; but a few legends have come down to us in abbreviated shape, which indicate that some then celebrated attempts and failures had taken place. No great faith can be attached to these legends, yet some of them are curious, if considered as the relation of attempts to sail upon the wind like soaring birds with rigid fixed surfaces.

Passing over as too scanty of record the myths of antiquity, perhaps the earliest legend of an experiment which we may fairly suppose to have been tried with an aeroplane is stated to be found in the somewhat fabulous chronicles of Britain,* wherein it is related that King *Bladud*, the father of King *Lear*, who is supposed to have reigned in Britain about the time of the founding of Rome, caused to be built an apparatus with which he sailed in the air above his chief city of Trinovante, but that, losing his balance, he fell upon a temple and was killed. This is about all there is of the legend, and as even that concerning King *Lear*, which Shakespeare worked up into his tragedy, has been suspected of being a myth, it is difficult to comment intelligently upon such a tradition ; yet it is not impossible that King *Bladud* (who was reputed to be a wizard, as were all investigators in ancient times), should have attempted to imitate the ways of the eagle in the air, and should have succeeded in being raised by the wind, when, for lack of the balancing science of the bird, he should have lost his equilibrium, and with a shear, a plunge, or a whirl have come in disaster to the ground.

A better authenticated legend seems to be that of *Simon the Magician*, who, in the thirteenth year of the reign of the Emperor *Nero* (about 67 A.D.), undertook to rise tow-

---

* Bescherelle, Histoire des Ballons, 1852.

ard heaven like a bird in the presence of everybody.* The legend relates that "the people assembled to view so extraordinary a phenomenon and *Simon* rose into the air *through the assistance of the demons* in the presence of an enormous crowd. But that St. Peter, having offered up a prayer, the action of the demons ceased, and the magician was crushed in the fall and perished instantly."

"It seems, therefore, certain" (adds M. *de Graffigny*) "from this tale, which has come down to us without any material alteration, that even in that barbarous age a man succeeded in rising into the air from the earth by some means which have unfortunately remained unknown."

The writer has seen the feat performed by soaring birds many times. He has seen a gull, standing upon a pile-head within 20 ft. of him, float up into the air without flapping, by simply facing the wind, opening his wings to their full extent, and keeping them rigidly extended to a sea breeze blowing at the rate of 14.40 measured miles per hour. The gull rose vertically about 2½ ft. above the pile-head, then drifted back about 5 ft., still rising slightly, when he altered by a trifle his angle of incidence, advanced against the wind, losing a little height, and was thenceforth in full soaring activity. Many other writers have seen the same kind of performance, including the still more difficult feat seen by M. *Mouillard*,† who observed in Africa an eagle spring from the top of an ash-tree, and without a single flap first descend from 7 to 10 ft., going against the wind, and upon this freshening to a squall, rise directly and slowly some 300 ft. into the air, while *advancing against the wind* some 150 ft. at the same time.

The reader may be further interested by the account of a somewhat similar feat, published in *L'Aéronaute* of October, 1890, by Mr. *Charles Weyher*, and which he describes as follows:

"One day when I was close to the Aqueduct of *Buc*, and the wind was blowing strongly down the valley, and therefore at right angles to the aqueduct, I saw a sparrow hawk come out of a hole marked *A* (fig. 36) on the sketch,‡ near the top, and on the leeward side.

"The bird left his hole and dove downward, his wings scarcely opened, and thus reached like a dart a point about the center of the opening of one of the arches. At this moment, when at *B*, he stretched his wings wide open and began circling, continued his orbits, drifting with the wind, until he attained an elevation of 800 to 1,000 ft. At this elevation, or the point

---

* Graffigny, La Navigation Aérienne, 1888.
† Mouillard, L'Empire de l'Air, 1881. Page 22.
‡ See page 79 for this sketch.

C, the sparrow hawk folded his wings almost completely and dove forward again upon a steep inclination, making use of the height gained to recover against the wind the distance which he had drifted, and to regain his hole, into which he entered gently, by simply opening his wings wide when within 7 to 10 ft. of the wall.

"It is well to observe that the bird in taking this journey, both going and coming back, expended no muscular work whatever, save the utterly inappreciable exertion of opening and folding up his wings twice."

The legend of *Simon the Magician,* which has led to the above digression, is clearly of Christian origin, as evidenced by the intervention of St. Peter, who is supposed to have been martyred in Rome about A D. 64. It is not known to be confirmed by any Roman record, such records having been largely destroyed during the dark ages ; but if the tradition be founded upon a fact, we may suppose *Simon,* after some preliminary trials, to have attempted to imitate, with a fixed aeroplane, in public, some of the evolutions of a soaring bird, and being unable to perform skillfully the necessary manœuvres, to have lost his equilibrium and his life.

There is another monkish tradition of the eleventh century concerning *Oliver of Malmesbury,* who in some of the accounts is styled " Elmerus de Malemaria," and who was an English Benedictine monk, said to have been a deep student of mathematics and of astrology, thereby earning the reputation of a wizard. The legend relates * that " having manufactured some wings, modeled after the description that *Ovid* has given of those of *Dedalus,* and having fastened them to his hands, he sprang from the top of a tower against the wind. He succeeded in sailing a distance of 125 paces ; but either through the impetuosity or whirling of the wind, or through nervousness resulting from his audacious enterprise, he fell to the earth and broke his legs. Henceforth he dragged a miserable, languishing existence (he died in 1060), attributing his misfortune to his having failed to attach a tail to his feet."

Commentators have generally made merry over this last remark, but in point of fact it was probably pretty near the truth. To perform the manœuvre described, of gliding downward against the breeze, utilizing both gravity and the wind, *Oliver of Malmesbury* must have employed an apparatus somewhat resembling the attitude of a gliding bird, but being unable to balance himself fore and aft, as does the bird by slight movements of his wings, head and legs, he would have needed even an ampler tail than the

---

* Bescherelle, Histoire des Ballons.

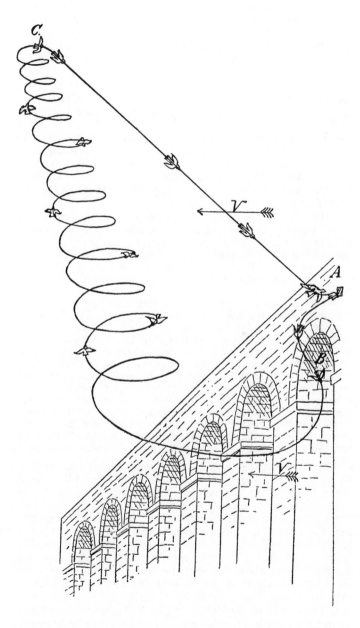

Fig. 36.—THE SPARROW-HAWK'S EXCURSION.

bird spreads on such occasions in order to maintain his
equilibrium. He would have failed of true flight in any
event, but he might have come down in safety.

A more explicit tradition of the same kind comes from
Constantinople, where, under the reign of the Emperor
Manuel Comnenus, probably about the year 1178, a *Saracen* (reputed to be a magician of course), whose name is
not given, undertook to sail into the air from the top of the
tower of the Hippodrome in the presence of the Emperor.

The quaint description of this attempt, as taken from
the history of Constantinople by *Cousin*, and given both
by *Graffigny* and by *Bescherelle*, so clearly describes an
aeroplane as distinguished from movable wings, and so
well indicates the difficulty of obtaining and maintaining a
proper balance with such an apparatus, that it is worth
quoting :

" He stood upright, clothed in a white robe, very long and very
wide, whose folds, stiffened by willow wands, were to serve as
sails to receive the wind. All the spectators kept their eyes in-
tently fixed upon him, and many cried, ' Fly, fly, O Saracen !
do not keep us so long in suspense while thou art weighing the
wind !'—*i.e.*, adjusting the angle of incidence and the equilibrium of the machine.

" The Emperor, who was present, then attempted to dissuade
him from this vain and dangerous enterprise. The Sultan of
Turkey in Asia, who was then on a visit to Constantinople,
and who was also present at this experiment, halted between
dread and hope, wishing on the one hand for the Saracen's success, and apprehending on the other that he should shamefully
perish. *The Saracen kept extending his arms to catch the wind.*
At last, when he deemed it favorable, *he rose into the air like
a bird ;* but his flight was as unfortunate as that of *Icarus*, for
the weight of his body having more power to draw him downward than his artificial wings had to sustain him, he fell and
broke his bones, and such was his misfortune that instead of
sympathy there was only merriment over his misadventure."

This account seems to be given with such circumstance
as to preclude the idea that it is merely the idle tale of
some lover of the marvelous. We may, therefore, fairly
seek to draw some inferences therefrom, which have not
been heretofore mentioned by other writers. The first is
that the apparatus was some form of aeroplane, because it
is likened to a robe instead of a pair of wings, and also
because no mention whatever is made of any flapping
action. The only active exertion described on the part of
the operator is that of the adjustment of the apparatus to
the prevailing wind, implying that it was so adjustable
that the angle of incidence might be regulated to obtain
an ascending effect, and the center of pressure be brought
to coincide with the center of pressure to produce fore and

aft equilibrium. The second inference is that the force of the wind was the only motive power relied upon, and that the apparatus was not blown away, but rose upon the wind like the gull which has been already described. This being possibly an instance of that mysterious phenomenon of " Aspiration" which was alluded to at the beginning of this account of " Progress in Flying Machines," and which will be found further exemplified when an account is given of the various experiments of Captain *Le Bris.* The third inference is that the defect lay in the maintenance of the equilibrium. That the apparatus started off properly balanced, but that so soon as a change occurred in the conditions, perhaps an erroneous manœuvre on the part of the Saracen, or perhaps a gust of wind on one side, the aeroplane lost its balance, and disaster ensued.

Only brief allusion need be made in this discussion to the writings of *Roger Bacon*, the eminent philosopher of the thirteenth century (1214-94). He seems to have prophesied both the balloon and the flying machine, but not to have tried or related any experiments. His writings will be found noticed in some of the encyclopædias, and in Wise's " History and Practice of Aeronautics," the latter book containing, moreover, references to the traditions which have here been mentioned, as well as to others which have been omitted.

One of the most celebrated traditions of partial success with a flying machine refers to *J. B. Dante*, an Italian mathematician of Perugia, who toward the end of the fourteenth century seems to have succeeded in constructing a set of artificial wings with which he sailed over the neighboring lake of Trasimene.* We have no description of the apparatus, but this was presumably an aeroplane, soaring upon the wind, for we have seen abundantly that all experiments have failed with flapping wings, man not having the strength required to vibrate with sufficient rapidity a surface sufficient to carry his weight in the air. Moreover, there would be a stronger and steadier wind over a lake than over the land, and the selection of a sheet of water to experiment over was very happy, as it would furnish a yielding bed to fall into if anything went wrong, as is pretty certain to happen upon the first trials. A similar selection has been recommended by *D'Esterno* and by *Mouillard*, and cannot be too strongly urged upon any future inventor who desires to make similar experiments. With adequate extent of surfaces, and (if he goes up at all) some prudence as to the height to which he allows the wind to carry him, he can thus acquire some insight into the science of the

---

* Tissandier, La Navigation Aérienne. Bescherelle, Histoire des Ballons.

birds, with no greater danger than that of numerous duckings.

Whether *Dante* grew overbold with some preliminary
successes, or whether he was impatient to display his
achievement before his fellow-citizens and his sovereign,
he attempted to repeat the feat in Perugia, on the occasion of the marriage of *Bartholomew Alviano* with the sister of *Jean Paul Baglioni*. Starting from the top of the
highest tower in the city of Perugia, he sailed across the
public square and *balanced himself for a long time in the
air*, amid the acclamations of the multitude. Unfortunately the iron forging which managed his left wing suddenly broke, so that he fell upon Notre Dame Church and
had one leg broken. Upon his recovery he seems to have
given up further experiment, but went to teach mathematics at Venice, where he died of a fever before he had
reached forty years of age.

Granting the tradition to be true, the apparatus used by
Dante must have been more manageable than any of its
predecessors, for the accident is said to have been due to
a breakage instead of a loss of balance. The latter, however, must have been still deficient, or *Dante* would have
renewed his experiments with a stronger forging. He
may have reasoned, moreover, that as the wind does
not blow with the requisite speed every day, and he knew
of no sufficiently light motor to take its place, the use of a
soaring machine would be very limited ; but it is very
unfortunate that we should have no description of the machine and its mode of operation.

Two somewhat similar experiments are alluded to in
M. G. de la Landelle's " Aviation," published in 1863,
but are too briefly described to give much of an idea as to
the kind of apparatus employed ; he says :

Paul Guidotti, an artist-painter, sculptor and architect, who
was born in Lucca in 1569, constructed wings of whalebone
covered with feathers, and made use of them several times
with success. Determining to exhibit his discovery, he took
flight from an elevation, and sustained himself pretty well in
the air for a quarter of a mile, but soon becoming exhausted,
he fell upon a roof, and his thigh-bone was broken. . . .

I might also cite the article from the Malaga newspaper, the
Courier of Andalusia, which was republished in several French
journals in March, 1863, stating that a peasant of the neighborhood, named *Francisco Orujo*, was said to have sailed in the
air a distance of one league with artificial wings in less than
fifteen minutes ; but why multiply examples ? It is better to
deduce from these occurrences, some of which are abundantly
authenticated, useful conclusions concerning the insufficiency
of man's muscular power, and concerning the sustaining power
of an inclined plane.

The writer has been thus far unable to find in other publications fuller accounts of the last two experiments mentioned, but it is a significant fact that the greater number of the experimenters who are said by tradition to have actually succeeded in floating for a short distance on the air, were men living in warm climates, where the soaring varieties of birds are much more numerous and more easily observed than in variable and colder climates. This suggests the inference that these experimenters had been watching the soaring birds, sailing upon fixed wings in every direction, and endeavored to imitate their evolutions. With the aid of the wind they may have attained a glimmer of success, but they failed in every instance for lack of accurate knowledge of what constitutes the science of the birds. Elsewhere than in warm climates the soaring birds are so few, they so frequently have to resort to flapping, that those who have not seen them sailing about for hours upon fixed, extended wings, deny even the possibility of such a performance, and only think of wings as oscillating surfaces ; and so when, in 1842, *Henson* patented his flying machine, the proposal to obtain support from the fixed surfaces of an aeroplane was hailed by many as a new and happy idea.

A top view of *Henson's* apparatus is shown by fig. 37. It consisted of an aeroplane of canvas or oiled silk

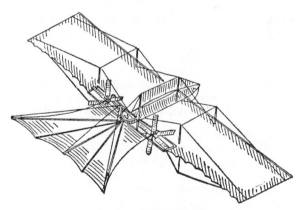

FIG. 37.—HENSON—1842.

stretched upon a frame made rigid by trussing, both above and below. Under this surface a car was to be attached containing a steam-engine, its supplies and the passengers. The apparatus was to be propelled by two rotating wheels, acting upon the air after the manner of a windmill. Back of these was a tail, also covered with canvas

or oiled silk, stretched upon a triangular frame, and
capable of being expanded or contracted at pleasure, or
moved up and down for the purpose of causing the
machine to ascend or to descend. Under the tail a rudder
was placed for steering the machine to the right or to the
left, and above the main aeroplane a sail or keel-cloth
was stretched, as shown, between the two masts which
rose from the car, in order to assist in maintaining the
course. The apparatus was to sail with its front edge a
little raised so as to obtain the required support or lift
from the air, and was to be started from the top of an
inclined plane, in descending which it was to attain a
velocity sufficient to sustain it in its further progress, the
steam-engine being only designed to overcome the head
resistance when in full flight.

Henson's patent indicates that he believed the correct
proportions to be about 2 sq. ft. of supporting surface to
each pound of weight, this being considerably in excess of
the proportion of the large soaring birds, and that the
motor required was about at the rate of 20 H.P. per ton
of weight. His general design evidences careful thought
and possesses some excellent features, but the form of his
aeroplane was crude and its equilibrium especially was
deficient. *Henson* stated in his patent :

The following are the dimensions of the machine I am making, and which will weigh about 3,000 lbs. The surface of the
planes on either side of the car will measure 4,500 sq. ft., and
the tail 1,500 more, with a steam-engine (high pressure) of 25
to 30 H.P.

Scaled from the patent drawings the intended dimensions of the main aeroplane appear to have been about
140 ft. long, in the direction of motion, by about 32 ft. in
width, this being considerably larger than the great
aeroplane that Mr. Maxim has been lately constructing in
England.

Henson did not realize his intention, for Mr. *F. W.
Brearey*, Honorary Secretary of the Aeronautical Society
of Great Britain, says in an article upon flying machines,
published in *Popular Science Review* in 1869,* in describing the *Henson* experiments :

The fact is the machine was never constructed ; for after
two abortive attempts to manufacture models at the Adelaide
Gallery, which should represent the dimensions before named,
he rejoined his friend (*Stringfellow*) at Chard, and the two
together commenced their experiments under a variety of
forms. . . . However, in 1844, they commenced the construction of a model ; *Henson* attending chiefly to the wood or
framework and *Stringfellow* to the power, and after many
trials adopted steam. This model, completed in 1845, meas-

---

* *Popular Science Review*, vol. 8, p. 1.

ured 20 ft. from tip to tip of wing, by 3½ ft. wide, giving 70 square feet sustaining surface in the wings, and about 10 ft. more in the tail. The weight of the entire machine was from 25 to 28 lbs. . . . An inclined plane was constructed, down which the machine was to glide, and it was so arranged that the power should be maintained by a steam engine, working two four bladed propellers each 3 ft. in diameter at the rate of 300 revolutions per minute.

A tent was erected upon the downs, 2 miles from Chard, and for seven weeks the two experimenters continued their labors. Not, however, without much annoyance from intruders. In the language of Mr. *Stringfellow:* " There stood our aerial *protégé* in all her purity—too delicate, too fragile, too beautiful for this rough world ; at least those were my ideas at the time, but little did I think how soon it was to be realized. I soon found, before I had time to introduce the spark, a drooping in the wings, a flagging in all the parts. ' In less than ten minutes the machine was saturated with wet from a deposit of dew, so that anything like a trial was impossible by night. I did not consider that we could get the silk tight and rigid enough. Indeed the framework was altogether too weak. The steam-engine was the best part. Our want of success was not for want of power or sustaining surface, but for want of proper adaptation of the means to the end of the various parts."

Many trials by day, down inclined wide rails, showed a faulty construction, and its lightness proved an obstacle to its successfully contending with the ground currents.

The above has been given verbatim, because of the importance of the experiments. Stated in plainer terms, it means that the machine was deficient in stable equilibrium for out-of-door experiments ; that " ground currents" or little puffs of wind would destroy the balance, and that in falling to the ground it would get more or less injured. That the experimenters, annoyed at the presence of spectators at these mishaps, endeavored to test their machine at night, with still less success, and finally gave it up in disgust. Mr. *Brearey* then continues :

Shortly after this *Henson* left England for America, and Mr. *Stringfellow,* far from discouraged, renewed his experiments alone. In 1846 he commenced a smaller model for indoor trial, and, although very imperfect, it was the most successful of his attempts (an illustration from a photograph is given) ; the sustaining planes were much like the wings of a bird. They were 10 ft. from tip to tip, feathered at the back edge, and curved a little on the under side. The plane was 2 ft. across at the widest part ; sustaining surface, 17 sq. ft.; and the propellers were 16 in. in diameter, with four blades occupying three-quarters of the area of the circumference, set at an angle of 60°. The cylinder of the steam-engine was ¾ in. diameter ; length of stroke, 2 in.; bevel gear on crank-shaft, giving 3 revolutions of the propeller to 1 of the engine. The

weight of the entire model and engine was 6 lbs., and with
water and fuel it did not exceed 6½ lbs.

The room which he had available for the experiments did
not measure above 22 yds. in length, and was rather contracted
in height, so that he was obliged to keep his starting wires very
low. He found, however, upon putting his engine in motion
that in one-third the length of its run upon the extended wire,
the machine was enabled to sustain itself ; and upon reaching
the point of self-detachment it gradually rose until it reached
the farther end of the room, where there was a canvas fixed to
receive it. Frequently during these experiments it rose after
leaving the wire as much as 1 in 7.

Stringfellow then went to Cremorne Gardens with the two
models, but found the accommodations no better than at home.
It was found that the larger model (*Henson's*) would run well
upon the wire, but failed to support itself when liberated.
Owing to unfulfilled engagements as to room, Mr. *Stringfellow*
was preparing for departure, when a party of gentlemen, un-
connected with the gardens, begged to see an experiment, and
finding them able to appreciate his endeavors, he got up steam
pretty high and started the small model down the wire. When
it arrived at the spot where it should leave the wire, it appeared
to meet with some little obstruction and threatened to come to
the ground, but it soon recovered itself and darted off in as fair
a flight as it was possible to make, to a distance of about 40
yds., farther than which it could not proceed.

Having now *demonstrated the practicability of making a steam-
engine fly*, and finding nothing but a pecuniary loss and little
honor, this experimenter rested for a long time satisfied with
what he had effected.

It is evident that, taught by experience, Mr. *Stringfellow* had obtained greater stability in the smaller model.
The aeroplane was shaped like the wings of a bird,
slightly curved on the underside and feathered at the back
edge, so that the elastic yielding of the feathers might
automatically regulate the fore and aft stability, like the
back fold in the paper aeroplane which has been described ; but the equilibrium was still insufficient for
experiment out-of-doors, and the important problem of
safely coming down was not solved at all, for to prevent
breakage the apparatus had to be caught in a canvas
fixed to receive it.

The sparrow-hawk, whose excursion has been described
(fig. 36), solved this last problem by simply tilting himself
back and opening his wings wide so as to stop his headway by increased air resistance. This possibly might be
done with a full-sized apparatus mounted by an operator,
but was scarcely practicable in a small model. To miti-
gate this difficulty Mr. *Stringfellow* increased the sustaining surface, so that it was 2.61 sq. ft. per pound, and
therefore might act like a parachute, but this largely in-

creased the "drift," and required more power, so that water and fuel could only be provided for a very brief flight, and the machine cannot fairly be said to have "demonstrated the practicability of making a steam-engine fly."

Mr. *Stringfellow* took the matter up again in 1868, and made further experiments with a somewhat different apparatus, which will be described in due course.

The next proposal for an aeroplane was that of *Aubaud*, in 1851, which is shown in fig. 38. It provided for a number of supporting planes, above which rotating screws were to furnish ascending power, while vibrating wings were to propel. The car containing the motor was to be beneath the planes, and equipped with legs or tubes containing compressed air, in order to ease off the shocks which might be encountered in alighting.

M. *Aubaud* seems to have reasoned that in order to secure safety in coming down, it was necessary to arrange

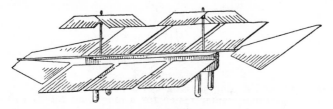

FIG. 38.—AUBAUD—1851.

matters so that the whole weight, or nearly the whole weight of the apparatus, could be sustained by screws when about alighting. This same general idea will be found to crop out in a number of subsequent proposals by inventors, who have believed that in order to come down safely it is necessary to design a machine which has enough power to start up by itself on level ground. This, of course, requires much more power than if only horizontal flight is provided for, and handicaps the inventor in an experimental machine.

The writer has been unable to ascertain whether *Aubaud* ever tested his apparatus experimentally. It seems clear that if he did, he must have become aware that no motor then known was sufficiently light in proportion to its energy to raise his machine into the air with screws, especially as he actually increased the ascending resistance by placing planes beneath the screws, so that the latter would not only have to sustain the weight, but also to overcome the vertical air pressure resulting from the movement. He advanced a meritorious proposal, however, by dividing the sustaining surface into several planes, an arrangement which we shall find (in describing Mr. *D. S. Brown's* experiments) to add materially to the stability ; but even with this feature the apparatus, as shown in the figure, is deficient in equilibrium, and would have come to grief many times if it had been experimented with.

The succeeding year *Michel Loup* proposed another form of aeroplane, which is shown both in plan and in

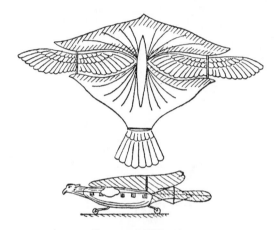

FIG. 39.—LOUP —1852.

side elevation in fig. 39. It is described both by M. *Dieu-aide* and M. *Tissandier* as consisting of a supporting plane propelled by four rotating wings, and provided with a rudder, also with legs beneath the car carrying wheels upon which the machine might roll upon alighting on the ground, an arrangement subsequently proposed again by many inventors.

The writer has been unable to find any record of experiments tried with this apparatus (which is chiefly here figured to show the wheeled feet), and it seems difficult to conceive of its successful operation.

In 1856 *Viscount Carlingford* patented both in England and France the aeroplane shown in fig. 40. and resembling in outline a falcon gliding downward with partially closed wings. In the center was a car or chariot, described by the inventor as follows :

The aerial chariot in form is something in the shape of a boat, extremely light, with one wheel in front and two behind, having two wings, slightly concave, fixed to its sides, and sus-

tainḙd by laths of a half hollow form pressing against them,
and communicating their pressure through the body of the
chariot from one wing to the other, and supported by cords,
whose force, acting on two hoops nearly of an oval shape,
hold the wings firmly in their position.

A tail can be raised and lowered at pleasure by means
of a cord.

The chariot is drawn forward by an "aerial screw"
in front thereof, "which screws into the air at an eleva-

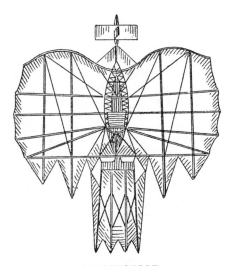

FIG. 40.—CARLINGFORD—1856.

tion of 45°, similar to the bird's wing ; and is turned by
means of a winch acting on three multiplying wheels."
This screw " is known as the Carlingford screw ; the blades
of this screw become more straight as they approach the
center, or, in other words, their edges become more direct
toward the center. . . . ' When a certain altitude is
attained the chariot may go several miles, perhaps 50 or
60, as it were, upon an inclined plane of air."

A novelty consisted in the mode proposed for starting
the chariot. It was proposed to suspend it by ropes between two poles, and then allow it (by drawing a trigger
suddenly) to fall upon the air and to be drawn "forward
with great velocity by the falling of the weight in front ;"
a method which we have seen to have been subsequently
adopted by M. *Trouvé* in starting his bird. If the invent-
or was thoroughly assured beforehand of the stability of
his apparatus at all angles of incidence, this would be an

elegant method of getting under way, but it would be somewhat awkward if there was any miscalculation about the position of the center of pressure. The writer has found no record of any experiments with the Carlingford apparatus.

The next apparatus to be noticed was thoroughly experimented with, and years were spent in the endeavor to put it into practical operation. It was first patented by M. *Felix du Temple*, a French naval officer, in 1857, and is shown in fig. 41 ; the top figure representing an end view from the rear and the lower figure being a top view. It consisted in two fixed wings of silk fabric, stretched by curved spars of wood or metal, and firmly attached to a car containing the motor. In front of this a screw was attached with a pivoted axle, in order to draw the apparatus forward. An horizontal tail hinged at the car was to regulate the angle of incidence and of flight of the machine, while a vertical rudder under the tail and separate from it was to steer to the right or left.

The car was to be shaped like a boat, lightly constructed of wood or iron ribs, and might be covered on the outside with tarred or rubber cloth. Beneath it were to be hinged three hollow legs, which might either be folded up or allowed to hang down. Strong springs inside of them were to carry rods or feet, at the outer end of which were to be wheels to roll over the ground. These legs were to be so adjusted in length that the apparatus should present an angle of incidence of about 20° to the horizon, and upon being put into forward motion, at the rate of about 20 miles per hour, by the screw, it was expected to rise upon the air and to enter upon its flight, the latter being regulated by the horizontal tail and by the vertical rudder.

The motor might be steam, electricity, or some other prime mover, and it was estimated that 6 H.P. would be required for an apparatus ; weighing one ton. This was a very great underestimate, for the proposed plan of driving the machine over the ground by means of the aerial screw would largely increase the resistance, and sufficient speed could not be obtained to rise upon the air.

M. *Du Temple* tried many experiments with models shaped like birds, and his patent indicates that he had carefully considered the question of stability, for he places the preponderance of weight toward the front of the car, provides for a diedral angle during flight by the flexibility and shape of the wings, and produces a slight turning up of the rear edge by making it flexible, all much as in the paper aeroplane which has been described. There was a weak point, however, in the fact that the center of gravity was not adjustable during flight, so as to correspond with the change in the center of pressure, produced by such alterations in the angle of incidence as might result from the action of the tail or otherwise.

When he began with the aid of his brother, M. *Louis du Temple*, to experiment on a large scale, the inadequacy

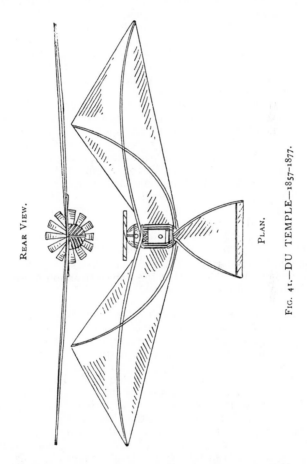

REAR VIEW. PLAN.

FIG. 41.—DU TEMPLE—1857-1877.

of all motors then known became apparent. They first tried steam at very high pressures, then a hot-air engine, and finally built and patented, in 1876, a very light steam boiler weighing from 39 to 44 lbs. to the horse power, which appears to have been the prototype of some of the light boilers which have since been constructed. It consisted in a series of very thin tubes less than ⅛ in. in internal diameter, through which water circulated very rapidly, and was flashed into steam by the surrounding flame.

This is understood to have been applied to a slightly modified form of the main apparatus, built in 1874 at Brest by M. *Du Temple.* This was calculated to carry a man, was 40 ft. across from tip to tip, weighed about 160 lbs., and cost upward of $6,000, the workmanship being very fine.

Careful search by the writer through various French and English publications has failed to discover any account of the operation of this machine, save the statement of M. *Tissandier* that "notwithstanding most persevering efforts, no practical results could be obtained in experimenting with this apparatus."

In 1858 *Jullien*, a French clock repairer of Villejuif, who had already exhibited in 1850 the first model of a fish-shaped navigable balloon which operated with its own motor, propeller, and steering gear, endeavored to prove what could be accomplished with aeroplanes. He exhibited to the French Society of Encouragement for Aviation a model weighing only 1 ¼ oz., although its aeroplane measured 39 in. across the line of motion. It was propelled by two screws each with two arms, and the power was furnished by the tension of a rubber band wound over two conical spools of equal diameter, like the *fusée* of a watch, in order to maintain the force uniform. The angle of incidence was about 10°, and the apparatus proceeded horizontally in a straight line a distance of 40 ft. in five seconds, with an expenditure of 0.52 foot-pounds per second.

Jullien proposed to build a large apparatus upon the same principle, but he failed to obtain the requisite financial backing. He saw, clearly enough, that he must have a light motor, and he began to experiment with electricity, seeking chiefly a light primary battery. In 1866 he announced that he had succeeded in devising an electric motor and battery weighing at the rate of 82 lbs. per H. P. with which he expected to drive an aeroplane through the air during an entire day ; but he did not receive the encouragement or aid of capital, and this ingenious inventor, who had struggled all his life long with inadequate means, died miserably poor, in a hospital in 1877.

The singular machine patented in 1860 by Mr. *Smythies,* while quite impracticable in form, is here mentioned in order to make known two novel proposals—*i.e.,* the utilization of the aeroplanes as steam condensers and the proposed means for shifting the center of gravity while in flight. A top view is shown by fig. 42. The apparatus was to consist of extended plane surfaces in order to furnish the support, and to be propelled by the alternating motion of wings actuated by a boiler and engine of a peculiar kind.

The boiler was to be upright, its top view being indicated by the circles at the junction of the two planes. It was to be fitted with small vertical water-tubes, thickly placed in a "flame-chamber" heated by the combustion of some volatile fluid mixed with air. Back of the boiler an upright cylinder was to be placed to work the wings up and down, feathering motion being imparted to the vanes composing them (by compound levers), so that they

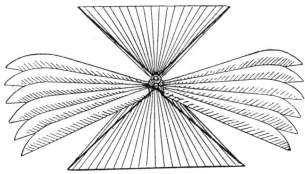

Fig. 42.—SMYTHIES—1860.

should separate slightly on the up-stroke and firmly close on the overlap on the down-stroke.

The whole of the two aeroplanes and of the upright boiler was to be encased in a closed flat bag of oiled silk or other light air-tight covering—that is to say, that a steam surface condenser was to be provideo by making a hood, tapering on its top and bottom from the thin edges at the front and rear of the apparatus to a thickness at the center equal to the height of the boiler. This hood or case was to be kept distended by spars and by light cords, and the spent steam was to be exhausted therein so as to be condensed by contact with the sides, and the water thus produced was to drain into a reservoir at the bottom, whence it was to be pumped back into the boiler, thus keeping the total weight of the apparatus constant and utilizing the same water over and over again.

The operator was to be suspended in an adjustable seat beneath the center of gravity of the machine, and by shifting his position sideways or fore and aft, was to guide the machine to the right or left, or up and down through the changes thus produced in the position of the center of gravity and consequent angle of incidence. Elastic legs beneath the whole were to break the fall on alighting ;

the descent being, moreover, retarded by the action of the wings, in which both the amplitude of movement and the overlap of the vanes or "feathers" were under the control of the operator through a series of cords.

This apparatus as designed was quite impracticable, but it indicates that the inventor had been watching the birds and had become aware of some requirements overlooked by his predecessors, such as the necessity for adjusting during flight the center of gravity to coincide with the center of pressure consequent upon changes in the angle of incidence, in order that the equilibrium may be preserved ; the necessity for more energetic efforts and greater angles of incidence at starting and in alighting than during horizontal flight, and also the necessity for an air condenser to liquefy the ; team, if the latter is to be used as a motive power, both to diminish the weight of the water required and to keep the weight constant.

In 1863 M. *de Louvrié*, a French engineer and mathematician, proposed the apparatus shown in fig. 43, which he called an "aeroscaphe." It consisted in a kite-like surface, stiffened by light cords to a mast above and to the car below, and capable of acting as a parachute, as well as of being folded like the wings of a bird. It was intended to

FIG. 43.—DE LOUVRIÉ—1863.

progress through the air either through the agency of a screw driven by a motor, or more directly by the reaction upon the air of some explosive substance, such as gunpowder. It was submitted to the French Academy of Sciences, but no experiments were made on a practical scale.

This proposal was the result of a mathematical investigation of the action of air upon aeroplanes and under the wings of birds, which was published by M. *de Louvrié*, first in *Les Mondes* and then in the *Aéronaute*,* wherein he showed that *Navier* and other French mathematicians had grossly overestimated the power required for flight. He also contended that the formulas then in current use for calculating the reactions of air upon oblique surfaces

---

* *L'Aéronaute*, September, October and November, 1868.

were erroneous, and he advanced the empirical formula of *Duchemin*, hereinbefore given, as agreeing much more closely with the observed facts. These writings of M. *de Louvrié* were sharply attacked by other aviators,* who had been promoting the imitation of flapping wings, and who denied altogether the possibility of soaring or sailing flight of birds, so that a lively controversy ensued, which has continued to this day, for M. *de Louvrié* published in 1890 † a new formula on a rational basis, of fluid action upon oblique surfaces, which agrees very closely in its results with those of the *Duchemin* formula, and he has also written an article on the "Theory of Sailing Flight," which is to appear shortly.

The "aeroscaphe" was practically a kite without string or tail, and its stability would probably have been found deficient, but M. *de Louvrié* tried in 1866 some experiments with a model weighing some 9 lbs. and exposing a surface of about 11 sq. ft. This ran upon a car on an inclined plane, and the "lift" and the "drift" were carefully measured by means of dynamometers. The results of these experiments demonstrated the fact that at all velocities and for all angles the "drift" or resistance of a plane to motion is to the "lift" or supporting pressure as the *sine* of the angle of incidence to its *cosine ;* as indeed was to be expected from theoretical considerations, and as had been laid down in 1809 by Sir George Cayley, in the article which has already been herein quoted.

As the *sine* is to the *cosine*, as the *tangent* of the angle M. *de Louvrié* deduced from his experiments the fundamental formula for the resistance of aeroplanes to be

$$R = W \ tang. \ @,$$

in which R is the resistance or "drift," W the weight, and @ the angle of incidence ; and calling L the "lift" and P' the normal pressure, at that angle of incidence, we may write further :

$$R = W \ tang. \ @ = P' \ sin. \ a$$
$$L = W \qquad = P' \ cos. \ a,$$

which formulas furnish at once the "drift" and the "lift" when either the normal pressure or the weight is known. In 1868 M. *de Louvrié* took out a second French patent for an aeroplane, in which he chiefly described the method of stretching the kite sail by adjustable radiating arms. It was to have a flat diedral angle to provide stability sideways, and was to be driven by a hot air engine, but no

---

* *L'Aéronaute*, December, 1868 ; January and February, 1869.
† *Revue de l'Aéronautique*, Vol. 3, page 102.

experiments seem to have been made. Since then he is understood to have abandoned the promotion of aeroplanes, and to expect more favorable results from his " anthropornis," which has already been noticed under the head of " wings and parachutes."

It may be incidentally here mentioned that a somewhat similar proposal for a circular kite or flat parachute was patented in the United States by Mr. *Wooton* in 1866. It must have been found, if experimented with, quite deficient in stable equilibrium ; any square, round, or polygonal surface being quite inferior in stability to the comparatively long and narrow wings which form the sustaining aero-planes of the birds.

While it seems certain, from the shape and arrangement of the apparatus of *Carlingford, Du Temple* and *De Louvrié*, that they had in mind the possibility of soaring upon the wind like a bird, they all proposed some kind of an artificial motor, and none of them was bold enough, in the face of existing prejudice, to propose to trust to the wind alone as a motive power. This was reserved for Count *D'Esterno*, who published in 1864 a very remarkable pamphlet * upon the evolutions and flight of birds, in which he gave the results of his many years of close observation, and formulated what he considered to be " the seven laws of flapping flight and the eight laws of soaring flight." He held that the act of flight involved the providing for three distinct and indispensable requirements— *i.e., equilibrium, guidance* and *impulsion*, and that the latter could be obtained from the wind whenever it blew strongly enough. He says in his pamphlet :

" Sailing flight labors under the disadvantage that it cannot take place without wind ; but, on the other hand, we can derive from the wind, when it blows, an unlimited power, and thus dispense with any artificial motor. In sailing (or soaring) flight, a man can handle an apparatus to carry 10 tons, just as well as one only carrying his own weight. Whoever has seen large birds of prey sailing upon the wind, knows that without one flap of their wings they direct themselves as they choose, save when they want to go dead with the wind or dead against it, on which occasions they must either tack or sweep in circles."

He patented in 1864 the apparatus shown in fig. 44, consisting of two wings hinged to a frame at the side of a central car, so that they could be set at any diedral angle, or even flapped should a motor be applied. The front end of the frame was provided with trunnions fitting in sockets insert-

---

* " Du Vol des Oiseaux," D'Esterno, 1864.

ed in the car, so that the rear end of the wings could be
raised or lowered, thus altering the angle of incidence,
and incidentally moving the wings forward or backward
with respect to the car. The wings themselves were to
be rigid within the triangles next to the car, and made
flexible in the rearward portion, where the curved ribs are
shown, which latter might be made of whalebone. It was
claimed that these wings would thus be capable of three
movements, corresponding to the first three "laws" laid
down by *D'Esterno*: 1. An up-and-down or flapping

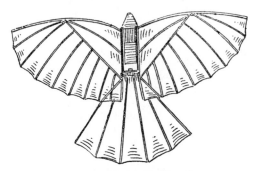

Fig. 44.—D'ESTERNO—1864.

action ; 2. A forward or backward inclination, and 3. A
torsional or twisting motion. The tail was connected with
the car with a universal joint, and had also three motions
corresponding to the next three laws—viz.: 4. An up-
and-down oscillation ; 5. A lateral displacement sideways,
and 6. Torsion. The car was provided with a movable
seat, and the operator could either sit thereon, and
shift his weight forward and back or sideways, or he
might stand up and effect the same object by swaying his
body or moving about, this action displacing the center of
gravity of the apparatus, and corresponding to the seventh
law of flapping flight, or means for the maintenance of
equilibrium ; to which for sailing flight *D'Esterno* added
still an eighth requirement or "law" in affirming the necessity for an initial headway.

He indicates in his book that an apparatus for one man
would weigh, with the operator, approximately 330 lbs.,
and spread an equal number of square feet of horizontal
surface, 215 sq. ft. of which would be in the two wings,
each being approximately 15½ ft. long by 7 ft. wide, and
he describes in his patent, mechanism somewhat crude,
chiefly consisting of ropes and drums, for producing the
various motions described.

The proposed mode of actual operation is not described, but it must have been nearly identical with the evolutions of the sparrow-hawk in the excursion which has been described. The apparatus would first descend in order to obtain an initial velocity, after which, having a speed of its own, it might utilize that of the wind. During this descent the fore-and-aft plane of the wings would have to point below the horizon, and if the reader will refer to the various attitudes of the sparrow-hawk on fig. 36, he will note that between points A and B his wings were scarcely open, in order to diminish the " drift." Once the speed gained, the apparatus would needs alter its center of gravity in order to change upward the angle of incidence and to come on a nearly level keel. If it went dead with the wind, its relative speed would have to be great enough to furnish support ; if going dead against the wind, it would lose headway but gain elevation ; and it might tack on a quartering wind or describe a series of helical orbits, like those of the birds. If the latter were chosen, the apparatus would, when it had the wind in its quarter, be sailing into the " eye of the wind" and faster than it blew, just as in the case of the ice boat ; while it would probably need to descend a little on the part of the circle when it was going with the wind, and would be enabled to rise materially when, upon facing the wind, the force of the latter was added to its own initial velocity. Thus, at every turn height would be gained, this being acquired when going against the wind ; and height once obtained, the apparatus would be able to sail in any direction by descending.

It will be noted, however, how many delicate manœuvres are requisite to accomplish these evolutions : to alter the angle of incidence, the direction, the speed, and to maintain the equilibrium at all times. The bird does all this by instinct ; he performs the exact manœuver required accurately, at exactly the right time, and he is always master of his apparatus ; but the man would be at a terrible disadvantage, his perceptions and his operations would be too slow, and a single false movement might be fatal.

There probably would have been many mishaps at first with Count $D'Esterno's$ apparatus had it been experimented, and being aware of this difficulty, he proposed that the experiments should be conducted over water sufficiently deep to break the fall, the apparatus being raised, like a kite, by a cord fastened ashore, which the operator could hold fast or abandon at will, while he was acquiring the science of the birds.

At that time (1864) aviation was not in public favor, and the very existence of soaring or sailing flight of birds was strenuously denied. It was held that there must be some small movements of the wings, which sustained the bird in the air, and which the observers had failed to detect, and it was not till the subsequent observations of *Pénaud, Wenham, Basté, Peal, Darwin, Mouillard*, and others that it was admitted that a bird might sail by the sole force of the wind without flapping.

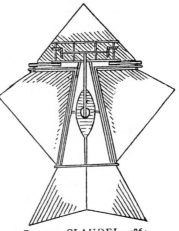

FIG. 45.—CLAUDEL—1864.

Count *D'Esterno's* proposal was generally laughed at as an evidence of mild lunacy, and whether because of this reason or because he distrusted the efficacy of his own mechanism, he did not build his apparatus. This is the more to be regretted because, being in possession of an ample fortune, he might have tried a number of valuable experiments which, if they did not result in success (as they probably would not), might yet have greatly advanced the fund of knowledge upon this intricate subject. Later on, when aviation grew in favor, and he was urged by members of the French Aeronautical Society, he conferred with various ingenious mechanics, and in 1883 he made an arrangement with M. *Jobert* to build a soaring apparatus. This was well under way when, in that same year, Count *D'Esterno* died at the age of 77. The apparatus was never completed, and, such as it is, has been deposited in the Exposition Aéronautique.

A singular proposal was that of M. *Claudel*, who patented in 1864 the apparatus shown in fig. 45. This consisted of two aeroplanes, one at the front and one back, propelled by two wings, lozenge-shaped, and rotated upon pivots at each apex. If they were made flexible the resistance of the air would bend these wings into an elongated screw, and some propelling effect might be produced. They were to be driven by bevel gears set in motion by a steam engine in the car ; but it is not known whether any practical experiments were ever tried.

In 1866 Mr. *F. H. Wenham* patented the meritorious proposal of superposing planes or surfaces above each other, so as to increase the supporting area without increasing the leverage. These were to be "kept in parallel planes by means of cords, or rods, or webs of woven fabric. . . . The long edges of the surface," made of silk or other light material, to be placed "foremost in the direction of motion." This system of surfaces being arranged above a "suitable structure for containing the motive power." If manual power was employed, the body of the operator was to be placed in a *horizontal* direction, and "the arms or legs to work a slide or treadle from which the connecting cords convey a reciprocating motion to oars or propellers, which are hinged above the back of the person working them."

In a very able paper, which has become classical, read at the first meeting of the Aeronautical Society of Great

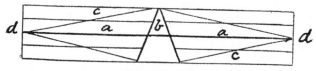

FIG. 46.—WENHAM—1866.

Britain, in 1866, Mr. *Wenham* gave an account of his observations, concluding with a very valuable discussion of the problem of flight, and with the following description of his experiments :

Having remarked how thin a stratum of air is displaced beneath the wings of a bird in rapid flight, it follows that in order to obtain the necessary *length* of plane for supporting heavy weights, the surfaces may be superposed or placed in parallel rows, with an interval between them. A dozen pelicans may fly, one above the other, without mutual impediment, as if framed together ; and it is thus shown how two hundred weight may be supported in a transverse distance of only 10 ft.

In order to test this idea, six bands of stiff paper 3 ft. long and 3 in. wide were stretched at a slight upward angle in a light rectangular frame, with an interval of 3 in. between them, the arrangement resembling an open Venetian blind. When this was held against a breeze, the lifting power was very great ; and even by running with it in a calm it required much force to keep it down. The success of this model led to the construction of one of a sufficient size to carry the weight of a man. Fig. 46 represents the arrangement, being an end elevation ; *a a* is a thin plank tapered at the outer ends, and attached at the base to a triangle, *b*, made of similar plank for the insertion of the body. The boards *a a* were trussed with thin bands of iron *c c*, and at the ends were vertical rods *d d*. Between these were stretched five bands of holland 15 in. broad and 16 ft. long. the total length of the web being 80 ft. (100 sq. ft. of surface). This was taken out after dark into a wet piece of meadowland one November evening, during a strong breeze, wherein it became quite unmanageable. The wind acting upon the already tightly stretched webs, their united pull caused the central boards to bend considerably, with a twisting, vibratory motion. During a lull, the head and shoulders were inserted in the triangle, with the chest resting on the base board. A sudden gust caught up the experimenter, who was carried some distance from the ground, and the affair, falling over sideways, broke up the right-hand set of webs.

In all new machines we gain experience by repeated failures, which frequently form the stepping-stones to ultimate success. The rude contrivance just described (which was but the work of

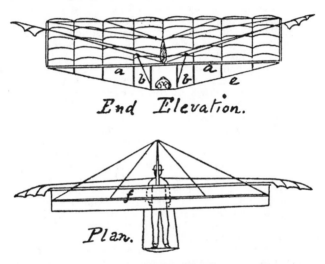

End Elevation.

Plan.

FIG. 47.—WENHAM—1866.

a few hours) had taught, first, that the webs or aeroplanes must not be distended in a frame, as this must of necessity be strong and heavy to withstand their combined tension ; second, that the planes must be made so as either to furl or fold up for the sake of portability.

In order to meet these conditions, the following arrangement was afterward tried : *a a*, fig. 47, is the main spar, 16 ft. long, $\frac{1}{2}$ in. thick at the base, and tapered, both in breadth and thickness, to the end ; to this spar was fastened the panels *b b*, having a base board for the support of the body. Under this, and fastened to the end of the main spar, is a thin steel tie band, *e e*, with struts starting from the spar. This served as the foundation of the superposed aeroplanes, and, though very light, was found to be exceedingly strong ; for when the ends of the spar were placed upon supports, the middle bore the weight of the body without any strain or deflection ; and further, by a separation at the base-board, the spars could be folded back with a hinge to half their length. Above this were arranged the aeroplanes, consisting of six webs of thin holland 15 in. broad (giving 120 sq. ft. of supporting surface); these were kept in parallel planes by vertical divisions 2 ft. wide of the same fabric, so that when distended by a current of air, each two feet of web pulled in opposition to its neighbor; and finally, at the ends (which were sewn over laths), a pull due to only 2 ft. had to be counteracted, instead of the strain arising from the entire length, as in the former experiment. The end pull was sustained by vertical rods, sliding through loops on the transverse ones at the ends of the webs, the whole of which could fall flat on the spar till raised and distended by a breeze. The top was stretched by a lath, *f*, and the system kept vertical by stay-cords taken from a bowsprit carried out in front. All the front edges of the aeroplanes were stiffened by bands of crinoline steel. This series was for the supporting arrangement, being equivalent to a length of wing of 96 ft. Exterior to this two propellers were to be attached, turning on spindles just above the back. They are kept drawn up by a light spring, and pulled down by cords or chains running over pulleys in the panels *b b*, and fastened to the end of a swiveling cross-yoke sliding on the base-board. By working this crosspiece with the feet, motion will be communicated to the propellers, and by giving a longer stroke with one foot than the other, a greater extent of motion will be given to the corresponding propeller, thus enabling the machine to turn, just as oars are worked in a rowing boat. The propellers act on the same principle as the winds of a bird or bat; their ends being made of fabric stretched by elastic ribs, a simple waving motion up and down will give a strong forward impulse. In order to start, the legs are lowered beneath the base-board, and the experimenter must run against the wind.

An experiment recently made with this apparatus developed a cause of failure. The angle required for producing the requisite supporting power was found to be so small that the crinoline steel would not keep the front edges in tension. Some of them were borne downward, and more on one side than the other, by the operation of the wind, and this also produced a strong fluttering motion in the webs, destroying the integrity of their plane surfaces, and fatal to their proper action.

Another arrangement has since been constructed having laths sewn in both edges of the webs, which are kept permanently distended by cross stretchers. All these planes are hinged to a vertical central board, so as to fold back when the bottom ties are released, but the system is much heavier than the former one, and no experiments of any consequence have yet been tried with it.

It may be remarked that although a principle is here defined, yet considerable difficulty is experienced in carrying the theory into practice. When the wind approaches to 15 or 20 miles per hour, the lifting power of these arrangements is all that is requisite, and, by additional planes, can be increased to any extent ; but the capricious nature of the ground-currents is a perpetual source of trouble.

If Mr. *Wenham* tried any further experiments with his apparatus, he has not, to the writer's knowledge, published an account of the results.   They would be nearly certain to be unsatisfactory for want of stable equilibrium.   The *Wenham* aeroplane was even more unstable than that of the bird, and the latter is constantly in need of adjustment, to counteract the " ground currents" and the variations in speed and in the angle of incidence.   Moreover, the *horizontal* position selected by Mr. *Wenham* was most unfavorable because unnatural to man, in directing the movements of an apparatus ; so that as often as he might rise upon the wind, just so often he was sure to lose his balance and to come down with more or less violence. The two propellers described by him would of course have proved quite ineffective in sustaining the weight, because man's muscular power is quite insufficient to have worked them with a speed adequate to that purpose, but they might have served to direct the course, had the equilibrium of the apparatus been stable.

Indeed, the writer believes that the first care of the aviator who seeks to solve the problem of flight, must be to seek for some form of apparatus which shall be, if possible, more stable in equilibrium than the bird.   The latter is instinct with life ; he meets an emergency instantly. Man's apparatus will be inanimate, and should possess automatic stability.   Safety is the first requisite—safety in starting, in sailing, and in alighting, and the latter operation must be feasible almost everywhere without special preparation or appliances before the problem can be said to be fairly solved.   It will probably prove the most difficult detail to accomplish, but it does not seem impossible when we see the feat performed by the birds so many times every day.

Mr. *Wenham's* proposal to superpose planes to each other in order to obtain large supporting surfaces without increasing the leverage, and consequent weight of frame, will probably be found hereafter to be of great value. The French experimenter *Thibaut* found that when two equal surfaces were placed *one behind the other*, in the direction of fluid motion, the resistance more nearly equaled that of the two separate surfaces than might be supposed.   Thus for two square planes, placed at a distance apart equal to their parallel sides, so as to cover each other exactly. M. *Thibaut* found the resistance equal to 1.7 times that of one single surface.   When the hinder plane projected by 0.4 of its surface beyond the front plane, the resistance was 1.95 times that of the single sur-

face.   This diminished to 1.84 times, when it became o.9.
Beyond this it reached nearly twice the resistance.*

Professor *Langley* found in his experiments with superposed planes, 15 × 4 in., soaring at horizontal speed, that
"when the double pairs of planes are placed 4 in. apart
or more, they do not interfere with each other, and the
sustaining power is, therefore, sensibly double that of the
single pair of planes ; but when placed 2 in. apart, there
is a very perceptible diminution of sustaining power shown
in the higher velocity required for support and in the
greater rapidity of fall." †

We may hence conclude that there will result a material,
indeed a great advantage in superposing planes, provided
they are so spaced as not materially to interfere with each
other, and provided also that they are arranged so as to
afford a good equilibrium.

The writer of this record of "Progress in Flying
Machines" originally hesitated whether he should include therein the account of the experiments of Captain
Le Bris, which is about to follow.   Not because he
deemed it incredible in itself, nor because, if correctly
stated, the experiments were not most interesting and instructive, but because the only account of them which he
had been able to procure was contained in a novel, in
which the author, to make the book more attractive, had
mixed up a love story with the record of the aerial experiments, which combination, in the present state of disbelief, the writer feared might be too much for the credu-
lity of the reader.   It is true that the author of the novel‡
said that the account of the experiments was scrupulously
correct, and that in this, the principal object of the book,
he had endeavored to be very exact, even at the risk of
detracting from his hero.   It is also a fact that the *Aéronaute*,§ in reviewing the book, said :

Throughout the novel are to be found absolutely historical
data concerning the artificial bird of *Le Bris*, his experiments,
his partial success, his mischances, and his deplorable final
failure, the latter not through a radical defect, but through
lack of method, steadiness in thought, and attention to details.

But still the writer hesitated to reproduce this tale of
an ancient mariner.

Fortunately, after a year's seeking, he succeeded in
getting a copy of an historical book, now quite out of
print, by the same author,‖ which gives without any em-

---

* Derval, " Navigation aerienne," 1889, p. 185.
† Langley, "Experiments in Aerodynamics," p. 40.
‡ " Les Grandes Amours," G. de la Landelle, 1878, page 369.
§ *Aéronaute*, March, 1879, page 86.
‖ " Dans les Airs," G. de la Landelle, 1884, page 210.

bellishments an account of Captain *Le Bris's* experiments, and quite confirms that given in the novel, wherein it is said to have been related " with scrupulous exactness." From the historical work, therefore, of M. de la Landelle, supplemented by his novel, the following account has been compiled of what seems to have been a very remarkable series of experiments on " aspiration."

Captain *Le Bris* was a French mariner, who had in his younger days made several voyages around the Cape of Good Hope and Cape Horn, and whose imagination had been fired by the sight of the albatross, sporting in the tempest on rigid wings, and keeping up with the fleetest ships without exertion. He had killed one of these birds, and claimed to have observed a very remarkable phenomenon. In his own words, as quoted by M. de la Landelle :

I took the wing of the albatross and exposed it to the breeze ; and lo ! in spite of me it drew forward into the wind ; notwithstanding my resistance it tended to rise. Thus I had discovered the secret of the bird ! I comprehended the whole mystery of flight.

Possessed with an ardent imagination, he early became smitten with the design of building an artificial bird capable of carrying him, whose wings should be controlled by means of levers and by a system of rigging ; and when he returned to France, and had become the captain of a coasting vessel, sailing from Douarnenez (Finistère), where he was born, and where he had married, he designed and constructed with his own hands the artificial albatross shown in fig. 48.

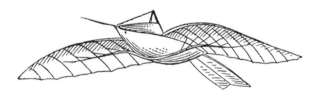

FIG. 48.—LE BRIS—1867.

This consisted of a body in the shape of a " sabot," or wooden shoe, the front portion being decked over, provided with two flexible wings and a tail. The body was built like a canoe, being 13½ ft. long and 4 ft. wide at its broadest point, made of light ash ribs well stayed, and covered on the outside with impermeable cloth, so it could float. A small inclined mast in front supported the pulleys and cords intended to work the wings. The latter were each 23 ft. long, so that the apparatus was 50 ft.

across, and spread about 215 sq. ft. of supporting surface ; the total weight, without the operator, being 92 lbs.   The tail was hinged so as to steer both up and down and sideways, the whole apparatus being, as near as might be, proportioned like the albatross.   The front edge of the wings was made of a flexible piece of wood, shaped like the front edge of the wing of the albatross, and to this, cross wands were fastened and covered with canton flannel, the flocculent side down.   An ingenious arrangement, which *Le Bris* called his *rotules* (knee pans), worked by two powerful levers, imparted a rotary motion to the front edge of the wings, and also permitted of their adjustment to various angles of incidence with the wind. *Le Bris* was to stand upright in the canoe (an excellent position), his hands on the levers and cords, and his feet on a pedal to work the tail.   His expectation was that, with a strong wind, he would rise into the air and reproduce all the evolutions of the soaring albatross, without any flapping whatever.

Le Bris's first experiment was conducted on a public road at Trefeuntec, near Douarnenez.   Believing, like Count *D'Esterno*, that it was necessary that the apparatus should have an initial velocity of its own, in addition to that of the wind, he chose a Sunday morning, when there was a good 10-knot breeze from the right direction, and setting his artificial albatross horizontally on a cart, he started down the road against the brisk wind, the cart being driven by a peasant.   The bird, with extended wings, 50 ft. across, was held down by a rope passing under the rails of the cart and terminating in a slip knot fastened to *Le Bris's* wrist, so that with one jerk he could loosen the attachment and allow the rope to run.   He stood upright in the canoe, unincumbered in his movements, his hands being on the levers and depressing the front edge of the wings, so that the wind should press upon the top only and hold them down, their position being, moreover, temporarily maintained by assistants walking along on each side.

When they came to the right turn in the road the assistants were directed to let go, and the driver was told to put his horse on a trot.   Then *Le Bris*, pressing on his levers, slowly raised the front edge of the wings to a very slight angle of incidence ; they fluttered a moment, and then took the wind like a sail on the under side, relieving the weight upon the cart so much that the horse began to gallop.   With one jerk *Le Bris* loosened the fastening rope, but lo ! it did not run, and the bird did not rise. Instead of this, its ascending power counterbalanced the weight of the cart, and the horse galloped as if at full liberty. It was afterward ascertained that the running rope had been caught on a concealed nail, and that the apparatus had remained firmly fastened to the cart. Finally the rails of the latter gave way, the machine rose into the air, and *Le Bris* said he found himself perfectly balanced, going up steadily to a height of nearly 300 ft., and sailing about twice that distance over the road.

But an accident had taken place. At the last moment the running rope had whipped and wound around the body of the driver, had lifted him from his seat, and carried him up into the air. He involuntarily performed the part of the tail of a kite ; his weight, by an extraordinary chance, just balancing the apparatus properly at the assumed angle of incidence, and with the strength of the brisk wind then blowing. Up above, in the machine, *Le Bris* felt himself well poised in the breeze, and exulted that he was about to pass two hours in the air ; but below, the driver was hanging on to the rope and howling with fright and anguish.

As soon as *Le Bris* became aware of this state of affairs, and this was doubtless in a very short time, he took measures to descend. He changed the angle of incidence of his wings, came down slowly, and manœuvred so well that the driver gently reached the soil, entirely unharmed, and ran off to çatch his horse, who had stopped when he again felt the weight of the cart behind him ; but the equilibrium of the artificial albatross was no longer the same, because part of the weight had been relieved, and *Le Bris* did not succeed in reascending. He managed with his levers to retard the descent, and came down entirely unhurt, but one wing struck the ground in advance of the other and was somewhat damaged.

This exploit naturally caused a great deal of local talk. Captain *Le Bris* was considered a visionary crank by most persons, and as a hero by others. He was poor, and had to earn his daily bread, so that it was some little time before, with the aid of some friends, he repaired his machine and was ready to try it again.

He determined this time to gain his initial velocity by dropping from a height, and for this purpose erected a mast, with a swinging yard, on the brink of a quarry, excavated in a sort of pocket, the bottom of which was well protected from the wind. In this quarry bottom he put his apparatus together, and standing in the canoe, it was suspended to a rope and hoisted up aloft to a height of some 30 ft. above the ridge, and nearly 100 ft. above the quarry bottom. A fresh breeze was blowing inland, and the yard was swung so that the apparatus should face both the wind and the quarry, while *Le Bris* adjusted his levers so that only the top surfaces of the wings should receive the wind. When he had, by trial, reached a proper balance, he raised upward the front edge of the wings, brought the tail into action through the pedal, and thought he felt himself well seated on the air, and, as it were, "aspired forward into the breeze." At this moment he tripped the hook suspending the apparatus, and the latter glided and sailed off toward the quarry.

Scarcely had it reached the middle of the pocket, when it met a stratum of wind inclined at a different angle from that prevailing at the starting-point—a vertical eddy, so to speak—probably caused by the reaction of the wind against the sides of the quarry. The apparatus then tilted forward ; *Le Bris* pressed on his levers to alter the plane of the wings, but he was not quick enough. The accounts of the bystanders were conflicting, but it was thought that the apparatus next oscillated upward, and then took a second downward dip, but in any event it finally pitched forward, and fell toward the bottom of the quarry.

As soon as the apparatus became sheltered from the wind it righted up, and fell nearly vertically ; but as it exposed rather less than 1 sq. ft. of surface to each pound of weight, it could scarcely act as a parachute, and it went down so violently that it was smashed all to pieces, and *Le Bris*, who at the last moment suspended himself to the mast of the canoe and sprang upward, nevertheless had a leg broken by the rebound of the levers.

This accident practically ruined him, and put an end for 12 or 13 years to any further attempt to prove the soundness of his theory. He had failed in both 'experiments for want of adequate equilibrium. He fairly provided for the transverse balance by making his wings flexible, but the longitudinal equilibrium was defective, as he could not adjust the fore-and-aft balance as instantly as the circumstances changed. The bird does this like a flash by instinct ; the man was compelled to reason it out, and he could not act quickly enough.

M. de la Landelle makes the following comment :

He lacked the science of Dante (of Perugia), but he was ingenious, persevering, and the most intrepid of men. He was entirely in the right in locating himself upright, both arms and legs quite free, in an apparatus which was besides exceedingly well designed. None was better fitted than he to succeed in *sailing flight* (*vol-à-voile*) in imitation of his model, the albatross.

In 1867 a public subscription at Brest enabled *Le Bris* to build a second artificial albatross, and this is the one represented by fig. 48. It was much like the first, but a trifle lighter, although a movable counterweight was added, intended to produce automatic equilibrium. The apparatus when completed was publicly exhibited, and attracted much attention ; but the inventor no longer had the audacity of youth, and he was influenced by number-less contradictory counsellors. He wanted to proceed as at Douarnenez, by giving an initial velocity to his appa-ratus, but he was dissuaded from this. He was also urged to test his machine with ballast, instead of riding in it himself, which at once changed all the conditions of equilibrium, as there was no longer command over a varying angle of incidence, and yet a first mischance led him to resort to the method of experimenting without riding in the machine.

M. de la Landelle relates the incident as follows :

Once only did he obtain something like an ascension, by starting from a light wagon, which was not in motion. He was on the levee of the port of commerce at Brest, the breeze was light, and the gathered public was impatient, through failure to realize that success depended wholly on the intensity of the wind. *Le Bris* was hoping for a gust which should enable him to rise ; he thought it had come, pulled on his levers, and thus threw his wings to the most favorable angle, but he only ascended a dozen yards, glided scarcely twice that dis-tance, and after this brief demonstration came gently to the ground without any jerk.

This negative result occasioned a good many hostile comments, and so the inventor no longer experimented in public ; but he had further bad luck ; the machine was several times capsized at starting, and more or less in-jured, being repaired at the cost of *Le Bris*, whose means were nearly exhausted. Then he tried it in ballast with varying success, and on one occasion, the breeze being just right, it rose up some 50 yds., with a light line attached, and *advanced against the wind* as if gliding over it. Very soon the line became slack, and the assisting sailors were greatly astonished, for the bird, without waver, thus proceeded some 200 yds.; but at the approach of some rising ground, which undoubtedly altered the direction of the aerial current, the bird, shielded from the wind, began settling down, without jolt, very gently, and alighted so lightly that the grass was scarcely bruised.

Encouraged by this partial success, *Le Bris* tried to reproduce the same results, but he met many mishaps, in which the apparatus was upset and injured. At last, one day, by a stiff northeast breeze, he installed his bird on top of the rising ground near which it had performed so well a few weeks before, and this time he meant to ride in the machine himself. He was dissuaded by his friends, and probably made a serious mistake in yielding to them, for the uncontrolled apparatus was not intended to adjust itself to a gusty wind.

At any rate, the empty machine rose, but it did not sustain itself in the air. It gave a twist, a glide, and a plunge, and pitched forward to the ground, where it was shattered all to bits. The wings were broken, the covering cloth of the canoe was rent to pieces, while the bowsprit in front was broken and forced back like a dart into the canoe.

The friends claimed that if the operator had been aboard, he surely would have been spitted and killed, but *Le Bris* maintained that if he had been aboard he could, with his levers, have changed the angle of the wings in time to avoid the wreck ; he blamed himself for having surrendered his better judgment, and he gave way to profound despair.

For this was the end. His second apparatus was smashed, his means and his credit were exhausted, his friends forsook him, and perhaps his own courage weakened, for he did not try again. He retired to his native place, where, after serving with honor in the war of 1870, he became a special constable, and was killed in 1872 by some ruffians whose enmity he had incurred.

Le Bris had made a very earnest, and, upon the whole, a fairly intelligent effort to compass sailing flight by imitating the birds. He finally failed for want of sufficient pecuniary backing, and also, perhaps, for lack of scientific methods and knowledge, for even at that day Captain *Béléguic*, a French naval officer, had called attention to the importance of securing longitudinal equilibrium, the lack of which caused the failure of poor *Le Bris*. Had he secured this he might have succeeded far better, especially if he had adhered to his original conception as to the necessity for that initial velocity against the wind, which served him so well upon the first trial and so ill upon the second. Singularly enough, he does not seem in all his subsequent experiments to have sought to give his apparatus that forward motion of its own, which he, like Count *D'Esterno*, originally held to be indispensable. He had also proposed to carry on his experiments at sea, from a steam vessel under way, and whatever may have been the cause that made him give up this idea, it was a misfortune, for his apparatus was capable of floating, he was himself an excellent swimmer, and had he experimented from a vessel under motion, not only would he have been safe, but he would have had no lack of wind to rise upon the air.

He seems, if the accounts given be correct, to have exhibited some very remarkable phenomena of " aspiration,'' which we shall find reproduced in one or two experiments yet to be noticed, and which the soaring birds exhibit every day to the observer, but he was baffled by the lack of fore-and-aft equilibrium.

In 1867 Mr. *Smyth* patented, in Great Britain, a combination of aeroplanes with lifting and driving screws, which is shown in end view by fig. 49. It consisted of a

FIG. 49.—SMYTH—1867.

cylindrical car with pointed ends, to carry the passengers and the motor ; of two aeroplanes, or light frames covered with silk, one at the front of the car and one at the rear, to furnish the supporting power when driven through the air by the propellers—one of which is shown in end view—and of two lifting screws, one above and the other below the car. These latter vanes were to be mounted at an adjustable angle upon vertical shafts passing through a tube in the car, the weight of the latter being transferred to them by means of a disk placed on rollers. This set of vanes was to be driven by one engine, and the horizontal propellers by another, so that the apparatus could be simultaneously lifted and driven forward if desired.

This arrangement is open to the objection that the resistance of the aeroplane has to be overcome in rising, and it is quite inferior to the proposal of *Crowell* patented in the United States in 1862, in which he had shown a pair of propellers (revolving in contrary directions), pivoted at the car, so that they might be brought overhead to ascend, and gradually dropped to the horizontal line to drive the apparatus forward, at which time a pair of aeroplanes, hinged at the car, and hanging vertically during the ascent, was also to be brought to a horizontal position to furnish a sustaining reaction.

Mr. *Smyth* seems to have apprehended that there might be some difficulty with his own arrangement, for he provides in his patent for a modification, in which he dispenses with the elevating propellers, and proposes to flap the sustaining surfaces or modified aeroplanes.

The principal feature of novelty, however, in Mr. *Smyth's* proposal was in the motor, which was to be a non-metallic apparatus, within which to explode gases to produce motion. This was to consist of a wooden cylinder, lined with a water-proof coating, and containing a series of india-rubber cells surrounded with water and connected with each other. Inside of these cells, which could be alternately collapsed or expanded, explosive mixtures of hydrogen and air were to be fired by electric sparks, the resulting expansion driving out a folding extension of the wooden cylinder, arranged like a concertina, and imparting motion to a jointed rod from which it was conveyed to the propellers.

The object was evidently to save the weight incumbent upon metallic engines, the patentee asserting that india-rubber cells are not injured by exploding mixed gases in them, so long as they are kept moist. It is probable that experiments were made with some models of this novel motor, but no account of them seems to have been published. Mr. *Smyth* had described his proposed machine at the 1867 meeting of the Aeronautical Society of Great Britain, but had no model in the exhibition of that society in the ensuing year, where, however, an analogous idea was brought forward by Mr. *D. S. Brown*, through a model in which the ordinary cylinder and piston of a steam-engine were replaced by an india-rubber cloth bag, the alternate inflation and expansion of which produced the stroke. It was stated that the first steam-engine constructed by Hancock ran on the common road with an engine of this description, but that as the process of vulcanizing india-rubber was not then known, the steam speedily softened the texture and escaped through the canvas.

Fig. 50 shows the form of aeroplane patented in Great Britain in 1867 by *Butler* and *Edwards*, and was evidently due to some recollection of their school-boy days, when

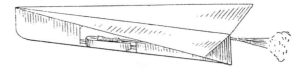

FIG. 50.—BUTLER & EDWARDS—1867.

they threw paper arrows in class. The stability of these little projectiles is quite good fore-and-aft, because the supporting surfaces increase in area while the intensity of the pressures diminish toward the rear, but the power required is great, and there is probably no aviating merit in this form. *Butler* and *Edwards* proposed to combine it in a variety of ways, superposing the sustaining planes, or placing two machines side by side, or both, and bracing between by diagonal ties. The form here shown is the simplest, the top planes being set at a slight diedral angle, in order to procure lateral stability.

The motive power was to be placed in a car, forward of the centre of figure, and capable of being moved forward and back, so as to shift the center of gravity to correspond with the center of pressure at varying angles of flight. The power was to consist in jets of steam issuing against the air in the rear ; but, suspecting that this would be enormously wasteful, the patentees reserved the right of using screw propellers, driven either by the reaction of jets of steam issuing from curved arms (Hero's æolipile) or by an ordinary steam-engine, in which case the steam was to be exhausted and condensed back into water, in cells formed by doubling the surfaces of the planes and thus providing hollow spaces.

We now come to a celebrated experiment, which attracted great attention at the time ; not so much because of the results obtained with the entire apparatus, for these were unsatisfactory, but because of the unprecedented lightness of the steam-engine in proportion to its estimated power.

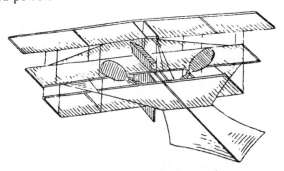

FIG. 51.—STRINGFELLOW—1868.

Mr. *Stringfellow*, whose experiments in connection with *Henson's* machine have already been noticed, was much impressed with Mr. *Wenham's* proposal for superposing planes, and when the Aeronautical Society of Great Britain advertised an exhibition and offered prizes, he constructed and entered two models for competition in 1868.

One of these models was the apparatus shown in fig. 51, and consisted of three superposed planes and of a tail, driven by a screw propeller moved by a steam-engine. The aggregate frontage was 21 lineal feet and the sustaining area of the three planes was 28 sq. ft., while the tail added some 8 ft. more, thus making some 36 sq. ft. The weight, including the aeroplane, engine, boiler, fuel and water, was under 12 lbs., thus giving a proportion of about 3 sq. ft. per pound, which was certainly ample for support. The engine was rated at one-third of a horse-power, but its weight is not stated.

The machine ran suspended along a wire, in the central transept of the Crystal Palace in London. It was forced by its propellers at great speed, but generally failed to lift itself from the wire, although Mr. *Stringfellow* said that it occasionally did so, and was then sustained by its superposed planes alone. This failure to rise may have been due to the fact, stated by M. Hureau de Villeneuve, that the axis of the screw was parallel with the sustaining planes, so that there was no angle of incidence, but a better explanation lies in the fact that the equilibrium was so imperfect that it was not safe, for fear of break-age, to liberate the machine. This, however, was done in the basement of the Crystal Palace, after the close of the public exhibition, a canvas being held to break the fall, and M. Brearey, an eye-witness, says :

When freed, it descended an incline with apparent lightness until caught in the canvas ; but the impression conveyed was that had there been sufficient fall, it would have recovered itself. . . . It was intended at the last to set this model free in the open country, when the requirements of the Exhibi-tion were satisfied, but it was found that the engine, which had done much work, required repairs. Many months afterward, in the presence of the author (M. Brearey), an experiment was tried in a field at Chard, by means of a wire stretched across it. The engine was fed with methylated spirits, and during some portion of its run under the wire, the draft occasioned thereby invariably extinguished the flames, and so these inter-esting trials were rendered abortive.

This apparatus (restored) is now in the National Museum of the Smithsonian Institution at Washington, where it is said that the engine cannot now develop anything like the power originally claimed.

The remark has well been made that so far as concerns the motive power, this apparatus ought to have flown. Its weight was only that of a goose, and it was said to have one-third of a horse-power. This may have been overestimated, but then Mr. *Maxim* estimates the power of a goose at 0.083 of a horse-power, and so there is ample margin.

There were two reasons for the failure. In the first place the sustaining surfaces were at least three times those of the goose, and hence the "drift," or horizontal component of the air pressure, would be much greater in the aggregate, so that the discrepancy in power is not as great as at first sight appears ; and in the second place. and more important, the equilibrium was so defective that the inventor did not dare liberate his model, while freedom of action is the first condition of successful experiment in flight. He had done better with single planes on the previous occasion which has been mentioned, but the failure with superposed planes in 1868 does not necessarily condemn them. It merely indicates that the surfaces must be of correct shape and skilfully arranged, and, if possible, produce automatic stability. If the equilibrium be unstable, like that of the bird, it is well enough to have more power than the goose, but it is much more important to possess its skill.

The second model of Mr. *Stringfellow* consisted in a steam-engine, similar to that applied to the complete apparatus, but larger. It was entered in the catalogue thus :

Light engine and machinery for aerial purposes, about half horse-power, cylinder 2 in. diameter, 3 in. stroke, generating surface of boiler, 3½ ft.; starts at 100 lbs. pressure in three minutes, works two propellers of 3 ft. diameter, about 300 revolutions per minute. With 3½ pints of water and 10 oz. of liquid fuel, works about ten minutes. Weight of engine, boiler, water, and fuel, 16¼ lbs.

Subsequently the examining jury estimated the power at one full horse-power, the weight of the engine and boiler alone being 13 lbs., and it awarded a prize of $500 to Mr. *Stringfellow* for "the lightest engine in proportion to its power, from whatever source that power may be derived."

With this prize money Mr. *Stringfellow* erected a building over 70 ft. long, in which to experiment with a view of ultimately constructing a large machine to carry a person to guide and conduct it, his experience with models having evidently impressed him with the necessity for intelligent control of any aerial apparatus not possessing automatic stability ; but he was already 69 years of age, his sight became impaired, and he died in 1883, without having accomplished any advance on his previous achievements.

A reference is made by M. de la Landelle in "Dans les airs" to an "Essay upon the Theory of Flight," published in 1869 at Copenhagen by Professor *Krarup-Hansen*, giving an account of an apparatus carrying a man who, by means of pedals, put into action horizontal wings whose action imparted motion to the machine, but the account is too scanty to determine whether this refers to beating wings or to an aeroplane.

Almost the same may be said concerning a large artificial bird, whose wings covered with feathers were more than 32 ft. across, constructed in 1845 by M. *Duchesnay* and exhibited in Paris, but which does not seem to have been tried in the open air. Somewhat more definite is the account given of some experiments by M. *Marc Seguin*, the celebrated French civil engineer, who in 1846 tried two series of experiments, one with lifting screws, and the other with an apparatus weighing 150 lbs., with a pair of canvas wings measuring 65 sq. ft. in area, from which he drew rather discouraging conclusions. He, however, tried another series of experiments in 1849, with an apparatus weighing 35 lbs. and worked by a man, the total weight, including ballast and operator being 232 lbs. This actually left the ground, and raised up some 6 to 8 in., but the effort required was evidently so violent as not to warrant further experimenting with man-power.

Fig. 52 exhibits an apparatus patented in 1871 by M. *Danjard*. This was to consist in parachute-like sails in front and at the rear, between which were to be placed two sustaining aeroplanes, between which again there was to be a pair of vibrating wings, which, in connection with

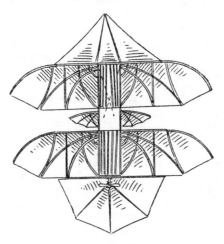

FIG. 52.—DANJARD—1871.

a screw, placed behind the car, were to furnish the impulsion. The front parachute was to be triangular in form, and made strong and rigid to cleave the air, while the rear parachute and the two aeroplanes were to be made flexible in the rear, so as to obtain a horizontal thrust from the escaping air compressed at the front. Under the rear parachute there was to be a rudder, to move to the right or to the left, and the machinery and aviators were to be in the central car. No motor is indicated save hand-power, but, of course, any primary motor could be applied if it were only light enough.

The apparatus is not known to have been experimented with. Probably M. *Danjard* dropped a lot of paper models and found that the arrangement of planes in pairs was more stable than a single aeroplane, and the figure is here given to show a combination which will be seen to have given fair stability in other experiments to be hereafter described.

The next experiment to be mentioned was important and quite successful upon the small scale on which it was tried. Fig. 53 represents an aeroplane with auto-

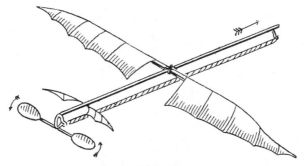

FIG. 53.—PÉNAUD—1871.

matic equilibrium, produced in 1871 by M. *A. Pénaud*, who called it his " planophore," and whose artificial bird and flying screw have already been noticed.

The motive power in this aeroplane was, as in his former models, the force of twisted india-rubber threads, fastened to a stick 20 in. long, and rotating a double-vaned screw 8 in. in diameter. The aeroplane, 18 in. across, by a width of 4 in., was fastened to the main stick at about its center, so that, through the leverage of the front end, the center of gravity of the apparatus should be slightly in front of the center of surface of the sustaining aeroplane. The outer ends of the latter were bent upward, so as to furnish lateral stability by a diedral angle, and the longitudinal stability was secured by fastening to the main stick back of the aeroplane, as shown, a small pair of wings or rudders, set at an angle of about 8° pointing below the horizon of the main aeroplane.

This was the important feature of the apparatus, and M. *Pénaud* not only showed experimentally that it furnished automatic equilibrium, but he also demonstrated*

---

* *Aéronaute*, January, 1872. Page 4.

the mathematical reasons why it should do so, in re-establishing, through the action of the air impinging upon this horizontal rudder set at a fixed angle, any deviation of the aeroplane from the horizontal line of flight. The principle is the same as that of the rear fold of the paper aeroplane which has already been described, and the following account of its mode of action was given by Mr. Bennett at the 1874 meeting of the Aeronautical Society of Great Britain :

The center of gravity of the machine is placed a little in front of the center of pressure of the aeroplane, so that it tends to make the model descend an incline ; but in so doing it lessens the angle of inclination of the aeroplane, and the speed is increased. At the same time the angle of the horizontal rudder is increased, and the pressure of the air on its upper surface causes it to descend ; but as the machine tends to turn round its center of gravity, the front part is raised and brought back to the horizontal position. If, owing to the momentum gained during the descent, the machine still tends upward, the angle of the plane is increased, and the speed decreased. The angle of the rudder from the horizontal being reduced, it no longer receives the pressure of air on its superior surface, the weight in front reasserts its power, and the machine descends. Thus, by the alternate action of the weight in front and the rudder behind the plane, the equilibrium is maintained. The machine during flight, owing to the above causes, describes a series of ascents and descents after the manner of a sparrow.

The weight of the entire apparatus was 0.56 oz., of which the rubber absorbed 0.17 oz., or about one-third. The surface was 0.53 sq. ft., so that the proportion was nearly at the rate of 15 sq. ft. per pound, and necessarily gave a slow flight.

The apparatus was publicly exhibited in August, 1871, to a group of members of the French Society of Aerial Navigation, in the garden of the Tuileries, and the model, guided horizontally by a small vertical rudder, not shown on the figure, flew several times in a circle, falling gently to the ground near its starting-point, when the power of the rubber was exhausted. The speed was not quite 12 ft. per second, or about the same as that of insects with the same relative surface in proportion to their weight, and the flight was 131 ft. in 11 seconds, with 240 turns of the rubber.

Subsequently M. *Pénaud* measured the power consumed in a very ingenious way. He found that with 60 turns of the rubber the apparatus would just hold its own —*i.e.*, hover in the same spot, against a wind of 9 ft. per second, and knowing the speed of rotation of the screw, as well as the weight of the apparatus, he deduced the conclusion that the power expended was at the rate of one horse-power for each 81 lbs. of weight, although M. *Touche,* who has revised the calculations, makes it about three times this amount—a result, of course, quite inferior to those obtained by Professor Langley and by Mr. Maxim, because, perhaps, of the greater proportion of surface to weight.

M. *Pénaud* was a very ingenious man, and might have accomplished great things in aerial navigation had not his career been cut short prematurely. He was one of the few men who have taken up the subject in his youth, for it is a singular fact that most of the scientific students of this inchoate research are now men of middle age, perhaps past the dread of being considered mentally unsound, but no longer with the ardor and the daring of youth. M. *Pénaud,* however, began before he was 20 years old, by producing his flying screw. He had intended to enter the French Navy, but a painful hip disease had brought him to crutches, and left him no career but that of scientific studies. These he directed to aerial navigation, and during six or seven years of improved health he impetuously investigated and experimented upon the various phases of the problem. Not only did he produce the three forms of apparatus which have been described, almost the first which have practically worked, each flying upon a different principle and all produced by one man, but he took a very active part in the investigations promoted by the French Society for Aerial Navigation ; making a scientific balloon ascent, in which he was somewhat injured, designing a plane table for platting the course of balloons, a guide rope break, a delicate barometer, a balloon-valve, a kite without a tail, balanced in the same way as his aeroplane, a form of explosion engine, a programme for experiments on air resistances, one for investigation of flight by instantaneous photography, since carried out by Professor Marey, etc., etc., towering above his fellow-members in discussions, in a way which must have excited many jealousies ; and he also contributed a number of very valuable papers to the *Aéronaute,* in one of which he endeavored to account for the mystery of sailing flight by showing that ascending currents in the wind were not rare, and were quite sufficient to explain all the phenomena.

These labors finally culminated in his taking, in 1876 (in partnership with M. *Gauchot,* a clever mechanician, who had produced an artificial bird), a patent for the apparatus shown in fig. 54, which was to be of sufficient size to carry up two men.

It was to consist of an aeroplane somewhat in the form

of an ellipse, built of a light framework covered both at top and bottom with varnished silk, and stiffened by wire stays radiating from two short masts above and from the car below the aeroplane. The outer ends of the aeroplane were to be flexible, or to be set at a diedral angle, in order to produce lateral stability, and the rear portion was also to be flexible and to bend upward, to produce the longitudinal stability, this being, moreover, provided for by two horizontal rudders, side by side, hinged at the rear, so as to set themselves automatically at the angle required to produce fore-and-aft equilibrium, upon the principle developed in the "planophore." Under these balanced horizontal rudders a vertical rudder was to steer to the right or left. A car, in the shape of a light boat, was to be rigidly attached just under the aeroplane, the steersman standing or sitting at the bow, with his head just above the top of the aeroplane, and protected from the wind by a glass box. Movable legs with rollers and springs were to be let down to get a preliminary run on land, or to alight in a glancing direction.

The motion was to be obtained from two propellers, placed at the front edge of the aeroplane, and rotating in opposite directions ; the power to be furnished by a steamengine—although M. *Pénaud* said frankly that he knew of none in practical operation sufficiently light for his purpose. He believed it ought not to weigh more than 15 to 22 lbs. per horse-power, and hoped to get one constructed within those limits. The engine was so to be located in the car as to bring the center of gravity of the apparatus one-fifth of the distance back of the front edge, and all the steering was to be done by the helmsman through a single lever, which might either be pulled or pushed to work the horizontal rudders, or twisted to work the vertical rudder.

The sustaining surface of the aeroplane was to be proportioned at the rate of about 0.24 sq. ft. per pound of weight, the whole apparatus with two aviators was to weigh 2,640 lbs., and required an engine of 20 to 30 H.P. to fly through the air at 60 miles per hour, with an angle of incidence of 2°.

This apparatus, the result of several years of study by an able man who bestowed very careful thought thereon, was never built. The writer of this does not believe it would have succeeded if it had been experimented with, but valuable data might have been obtained. Aside from the difficulty about a light motive power, a difficulty now almost removed, it may be questioned whether the general form of the aeroplane was the best possible to glide upon the air, and whether the longitudinal equilibrium would have been as well preserved as in M. *Pénaud's* toy model.

If not, then a sustaining surface of only 0.24 sq. ft. per pound would have been exceedingly dangerous. The horizontal rudders, when left to adjust themselves, were expected to regulate the automatic balance, but they were also expected, when actuated by the steersman, to alter the angle of incidence, in order to cause the apparatus to rise or to fall. Such a change in the angle would necessarily alter the position of the center of pressure, and there was no provision for making a corresponding change in the center of gravity, other than by the displacement of the aviators themselves, or that of the fuel, water, or boiler, which displacement would be nearly impracticable.

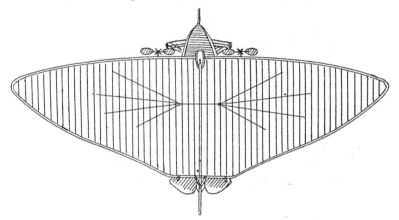

FIG. 54.—PÉNAUD & GAUCHOT—1876.

This was the weak point, for the pressure being then applied at a point differing from the center of gravity, would act with a leverage upon the apparatus and tilt it either forward or backward longitudinally, so that, had it been experimented with on a practical scale, it might have experienced a forward sheer or a plunge, either from too great an action of the horizontal rudders in rising or in coming down, or, as in the case of poor *Le Bris's* second experiment, from encounter with a stratum of wind of different horizontal direction than that for which the machine was adjusted.

Perhaps surmising the possibility of some such action, M. *Pénaud* suggested that the experiments should be conducted over a sheet of water. The apparatus might also have been suspended between two very high masts, or from a captive balloon, but probably the best results would have been obtained by experimenting entirely clear of any restraining supports.

At any rate no funds were forthcoming for the construction of the full-sized machine. M. *Pénaud* was criticised, decried, misrepresented, and all sorts of obstacles arose to prevent the testing of his project. He lost courage and hope, his health gave way, and he died in October, 1880, before he had reached 30 years of age.

He had doubtless done much toward solving the difficult problem of automatic stability in the air, but the French aviators do not seem to accept M. *Pénaud's* device as a solution of the question of longitudinal equilibrium. They claim that it consumes too much power in the constant readjustment of the stability, and that in a full-sized navigable apparatus it would not act quickly enough to prevent disaster. In September, 1890, M. *Hureau de Villeneuve* published a paper in the *Aéronaute*, in which he treats the problem of stability as yet to be solved, and suggests the inverted cone, as exemplified in the parachute of Cocking (which failed simply by reason of faulty construction), and he proposes as a possible solution of the equilibrium in both directions, that the aviating surface shall be made to conform to the development of an inverted cone. That is to say, that its lines, whatever they may be in the ground plan, shall in vertical projection follow the development of an inverted cone, placing the center of gravity so as to correspond with the position of the apex, and this he seems to have illustrated with a number of working models by cutting out various forms of birds out of an inverted cone.

Pénaud said that his solution was practically the same as that of Sir *George Cayley*, with whose labors he was not acquainted at the time that he hit upon the device for his " planophore." In his articles in *Nicholson's Journal*, published in 1809 and 1810, Sir *George Cayley* shows that the lateral stability is easily secured by placing the wings, either curved or plane, at a slight diedral angle to each other, and he lays down the principle, that in order to secure the longitudinal stability : 1. The center of gravity must be made to occupy a position directly under the center of pressure ; and 2. The aeroplane requires, to steady it, a rudder in a similar position to the tail in the bird. He then continues :

All these principles upon which the support, steadiness, elevation, depression, and steerage of vessels for aerial navigation depend have been abundantly verified by experiments, both upon a large and small scale. I made a machine having a surface of 300 sq. ft., which was accidentally broken before there was an opportunity of trying the effect of the propelling apparatus, but its steerage and steadiness were perfectly proved, and it would sail obliquely downward in any direction according to the set of the rudder. Its weight was 56 lbs., and it was loaded with 84 lbs., thus making a total of 140 lbs., or about 2 sq. ft. to 1 lb. Even in this state, when any person ran forward in it with his full speed, taking advantage of a gentle breeze in front, it would bear upward so strongly as scarcely to allow him to touch the ground, and would frequently lift him up and convey him several yards together. . . . It was beautiful to see this noble white bird sail majestically from the top of a hill to any given point of the plain below it, with perfect steadiness and safety, according to the set of its rudder, merely by its own weight, descending in an angle of about 8° with the horizon.

A number of very interesting experiments upon the stability of aeroplanes were tried in 1873 and 1874 by Mr. *D. S. Brown.* He had begun by seeking for a light motive power, and his proposal for a steam-engine with an india-rubber bag instead of a cylinder has already been noticed ; but becoming aware of the enormous importance of stable equilibrium, he turned his attention in that direction. He exhibited at the meeting of the Aeronautical Society of Great Britain, in 1873, a model consisting of two planes of equal size, one placed before the other at some distance and connected by a rod, which arrangement showed much greater stability than a single plane, and he followed this up at the next meeting, in 1874. by exhibiting models of what he called his " aero-bi-plane," which showed still further improvement in the stability, in consequence of " constructing the anterior edges or frames of the planes rigid, and the other parts yielding or elastic." The two planes might be rectangular and have their anterior edges straight, or these might be curved to diminish the air resistance, and the surfaces were placed one behind the other in the same general plane, so that they did not, as in the case of *Pénaud's* " planophore," make a slight horizontal angle with each other.

Mr. *Brown* stated that " the aeroplane should not be inclined to its path of motion, but its surface should form a direct line with it. . . . the plane being kept at the same elevation by slightly directing its course upward, sufficient to compensate for any fall which may take place." As, however, there was in the model a horizontal rudder, which regulated the angle of incidence in flight, Mr. *Moy* pointed out that the claim was a distinction without a difference. In one model the planes were connected with each other through double rods, so as to admit of a load, representing a car, being placed between them, and Mr. *Brown* stated that in this form it might be termed a progressive parachute, which was only supported by the air when in forward motion. When this motion was stopped—and this might be done by bringing it sud-

denly into a large angle with the horizon, so as to increase
the resisting surface—the model settled down to the floor.
It being, therefore, necessary that the apparatus should
start with an initial motion, this was given by an india-
rubber rope fastened at one end to a post, and at the other,
by means of a ring, to a vertical bolt inserted in the under
part of the bi-plane, so that it might be released when the
rope slackened.   The experiments are described in the
report of the meeting as follows :

Mr. *Brown* launched several planes of different dimensions.
All showed perfect stability, and, save one or two, floated in
the air in a horizontal position across the room, a distance of
between 20 and 30 ft., and, apparently, in some instances could
have gone further without falling had not the walls intervened.
One he suddenly pressed downward in a perpendicular direction
by striking it with a stick when in the air ; this caused it to dart
forward with great velocity in a horizontal course.   Mr. *Brown*
considered this an illustration of true flight, as the planes were
only inclined the moment he struck the connecting-rod.   During the flight they recovered their horizontal position and offered
no resistance to the air.

It may be noticed that this arrangement of surfaces,
which Mr. *Brown* referred to as " first steps to flight,"
differs from that of M. *Pénaud* in making the rear plane
of the same size as that of the front, and parallel therewith, the automatic stability being obtained through the
flexibility of the posterior edge, which acts much in the
same way as the upward inclination of the rear plane or
rudder in *Pénaud's* apparatus.   Whether either of these
arrangements, thus slightly differing in construction, will
prove adequate in practical operation upon a large scale,
can only be ascertained by experiment, but it may be
stated that the British aviators have not accepted Mr.
Brown's proposal as a solution of the problem of equilibrium, and that some of them believe that he made a mistake in placing his planes parallel with each other ; " no
change of action taking place whether the planes move
from the horizontal to 45° or to any other angle."

The next apparatus to be noticed will be found described in most of the articles on flight in magazines and
in encyclopædias, but the writer of these lines has been
fortunate enough to obtain from Mr. *Moy* himself still
further particulars concerning an experiment which has
well been characterized in the reports of the Aeronautical
Society of Great Britain as " one of the most determined
attempts at solving the problem which has yet taken
place."

Fig. 55 shows a front view, from a photograph, of
" Thomas Moy's aerial steamer," which was tried in the

open air at the Crystal Palace near London in June, 1875. The supporting surfaces consisted in two aeroplanes, one in front and the other behind the propelling aerial wheels; the planes being of linen, stretched upon bamboo canes, and set at an angle of 10° with the horizon, the rear plane being placed higher than the front plane, but parallel therewith. A third steering plane, of smaller size, governed by a horizontal wind wheel with screw vanes, was placed in the rear to serve as a horizontal rudder. The front plane measured 50 sq. ft., and the after plane had 64 sq. ft. of surface; their true size in relation to the whole apparatus being inadequately shown in the figure, because of the perspective, as they are seen nearly edgewise.

Between the two supporting aeroplanes were placed two propelling aerial wheels 6 ft. in diameter, each provided

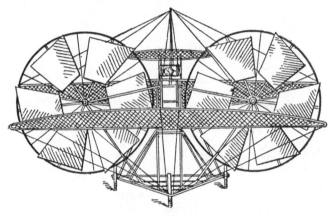

FIG. 55.—MOY—1875.

with six blades. These were first made of thin laths to approximate to true helices, but were afterward made of Scotch cambric. The blades or vanes were by a most ingenious and simple arrangement caused to change their angle of incidence as they rotated, so as to be "successively caused to be inclined to the line of onward motion of the machine, in such a manner that the blades on one side of the neutral line will be caused to act downward on the air with both a raising and propelling effect, while those on the other side thereof will, in their upward course, be impinged upon by the air with only a lifting tendency." This being in effect an aerial screw in which the pitch was variable in every portion of the revolution, and constituting the chief feature of novelty in the whole apparatus.

The steam-engine was placed between the two aerial wheels, and was a marvel of lightness. The diameter of the cylinder was 2¼ in. and the stroke 3 in., with 520 to 550 revolutions per minute. The heating surface was 8 sq. ft. or 2⅔ sq. ft. per horse-power, the boiler being of the water-tube description, and the steam pressure was 120 to 160 lbs. per square inch, the fuel being liquid and burned in Russian lamps. The engine weighed, with the boiler, 80 lbs. and developed fully 3 H.P., being at the rate of 26⅔ lbs. per horse-power, or about the same as the 1868 engine of Mr. *Stringfellow* as rated by himself. Mr. *Shill*, a clever mechanician, who exhibited a remarkably light engine in 1868, was associated with Mr. *Moy* in producing the 1875 engine, and had an interest in the patents, so that the apparatus was also known as the "Moy & Shill aerial steamer." It was 14 ft. long and about 14 ft. wide, was mounted on three wheels, and weighed 216 lbs., thus being proportioned at the rate of 0.53 sq. ft. of sustaining surface per pound of weight, omitting the lifting effect of the aerial wheels, which measured 60 sq. ft. more.

The inventor estimated that at a speed of 35 miles per hour the apparatus would be able to rise from the ground and glide upon the air, and this estimate seems fully confirmed by Professor Langley's recent experiments, which show that the uplift on a plane surface of 114 sq. ft. at an angle of 10° would be fully 206 lbs., while somewhat higher results are obtained from the table of "lift" and "drift" heretofore given herein, when taken in connection with Smeaton's table of wind pressures.

After some preliminary tests a path around one of the fountains at the Crystal Palace was selected, which had a diameter of nearly 300 ft.; a pole was erected at the center of the fountain, and two cords were run from the top of the pole to each end of the machine, in order to keep it at a uniform distance from the center. The gravel had been rolled, and steam was got up. The gravel, however, proved too rough, it shook the steamer and largely increased the traction. Then a board walk was laid over the path, and again steam was got up and a good run was made around the fountain, the machine (which was only a large model and could not carry an engineer) being wholly propelled by the action of the aerial wheels upon the air, acting only as drivers.

The utmost speed attained was 12 miles per hour, while 35 miles an hour was required to cause it to leave the ground. This indicated that the resistances had been underestimated, which resistances consisted in the traction upon the boards, the air resistance on the framing, cordage and ground wheels, and also in the " drift" due to the inclination of the sustaining planes. With our present knowledge we can say that at a speed of 35 miles per hour (6 lbs. per sq. ft.) the latter would have been : $114 \times 6 \times 0.0585 = 40$ lbs., and as the speed would have been 3,080 ft. per minute, the power required by the " drift" was : $\dfrac{40 \times 3080}{33.000} = 3.73$ H.P., to say nothing of the other elements of resistance, so that it is not strange that only 12 miles an hour was attained.

Mr. *Moy* needlessly handicapped himself in starting from the ground by a level run. He reasoned, like many others before and since, that "when they were coming down power was wanted, and, of course, power was especially wanted when they were going up," but he encountered thereby, in an experimental machine, all the additional resistance of the traction upon the boards. He considered the propriety of launching the apparatus from a height, or down an incline, but then this costly machine, built wholly at his own expense, would surely have come to grief, for he says that "the transverse stability was better than the longitudinal stability, but both were bad," and unless this was first remedied, it really was not safe to experiment.

Mr. *Moy* also placed his sustaining aeroplanes at too obtuse an angle, for if he had simply doubled their area, and inclined them at 5° instead of 10°, the "lift" would have been, by the table, at 35 miles per hour : $228 \times 6 \times 0.173 = 236$ lbs., or practically the same as before, but the "drift" would have been diminished to : $228 \times 6 \times 0.0152 = 20.79$ lbs., or about one half of that heretofore calculated.

Such experiments would doubtless have been tried had ample means been forthcoming, but other things were more pressing, for it was recognized that some modifications would be required in the steam generator, which was provided with six Russian lamps burning methylated spirits, and it was found that when running in the open air, the fumes from the three forward lamps extinguished the three after lamps, and thus reduced the power one-half. Before even this difficulty could be remedied the machine was seriously injured by the wrecking of the bamboo aeroplane frames, while it was being moved *stern first* across the grounds, in a fierce gale, and Mr. *Moy* then decided to rearrange it for experiment, as to its vertical lifting power, by substituting 12-ft. aerial wheels with vertical axles tried under cover.

The total surface of these new aerial wheels was 160 sq. ft., and the weight, including engine, boiler and all accessories, was 186 lbs. It was found that by counter-balancing 66 lbs. with levers, the wheels would lift the remaining 120 lbs., thus showing a lift of 40 lbs. per H.P., or about the same as the best performance which has been attained with other forms of screws.

Mr. *Moy*, as the result of his various experiments, then proposed to build a much larger apparatus, with an engine of 100 H.P., and capable of carrying several men, both to avail of the diminished relative weight and resistance of larger engines and to secure intelligent control while in action ; but the money could not be secured for this purpose. It would necessarily have cost a large sum, and at that time (1875) not only was the whole subject of aerial navigation generally considered as visionary, but there was not sufficient knowledge to enable the public, or even scientific men, to distinguish the difference between a wild proposal, sure to fail to compass flight, and a promising experiment which was worth following up—a condition of affairs which has in a measure continued to this day, and which this account of " Progress in Flying Machines" is partly written to remedy.

So Mr. *Moy* got no money, but he had instead " two chancery suits about shares in his patents, with no help from any one ;" his experiments had brought him down in funds, and he had to turn to hard work to live. As he justly remarks, " Unless you can lift the last ounce of a model, the unscientific people call it a failure, and few can appreciate that as size and weight increase, the relative hull resistance decreases, by reason of its diminished surface in proportion to its cubic contents." He, however, continued to take an active interest in the subject ; read papers at the meetings of the Aeronautical Society of Great Britain, made private experiments with planes, both in air and in water, as well as with methods for securing automatic stability, and with an improved method of propulsion.

In 1879 Mr. *Moy* exhibited at the meeting of the Aeronautical Society the small flying model shown in fig. 56, which he described as a " military kite" mounted upon wheels and provided with propelling gear. The front plane measured 660 sq. in. of surface, and the after plane, of half its linear dimensions, measured 165 sq. in. They were made of cambric, fastened to a central box-girder of thin pine, running lengthways, and mounted on small wheels, the aeroplanes being given a diedral angle, as shown, and the angle of incidence fore and aft being adjustable. At each end of the central stick or box-girder cross-arms were fastened, which held in position strands of india-rubber strings, one on each side of the central stick and parallel

therewith, the untwisting of which rotated (in opposite
directions) two screw propellers with two vanes each.
These propellers are removed in the plan and side view

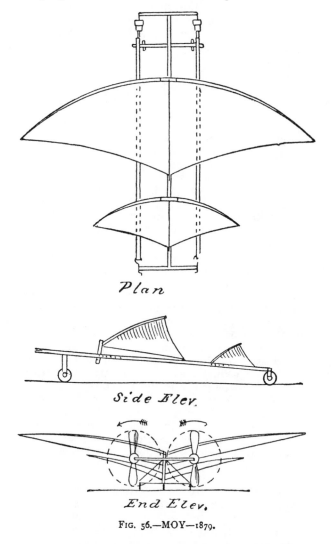

Plan

Side Elev.

End Elev.

Fig. 56.—MOY—1879.

of fig. 56, for the sake of clearness, but are shown in the
end elevation or front view.

The model weighed 24 oz., of which 3½ oz. was in the
india-rubber springs, and with 500 turns of the rubber it

would run on its wheels over a smooth surface, and under favorable circumstances would rise for a short distance upon the air. It would, of course, fly from the hand like *Pénaud's* planophore, but showed, like it, great waste of power, the supporting surfaces being at the rate of 3.82 sq. ft. per pound, and the angle of incidence required for it to rise from the ground being 8° ; and more unfavorable than a flatter angle, at which, however, it would not have possessed sufficient " lift." Mr. *Moy* says that " its transverse stability was very good, but its longitudinal stability was defective, and was a perfect puzzle at that time, but now all these troubles are overcome."

He has recently patented in England a method for automatically securing horizontal stability, and it is to be hoped that he will be enabled to renew his experiments.

Mr. *Moy* has also been an observer of soaring or " sailing" flight, and he described the sailing of the albatross in a wind, on rigid wings, in a paper read before the Aeronautical Society in 1869. Deeming it possible for man to imitate this performance, but recognizing the prodigious difficulty of reproducing the complicated shape and arrangement of curved surfaces (not planes), with which impulse is received from the wind, and of imitating the exquisite balancing through which the soaring birds perform this feat, he read a paper before the " Balloon Society," in 1884, in which he proposed an ingenious method of carrying on the many experiments required, with no greater danger than that of getting wet, by taking a start at sea from a lifeboat on the crest of the waves in a gale, equipped with a pair of wings and an inflated Boynton dress. In the hope that some aviator, favorably situated, may try this experiment, the paper, rewritten for the purpose, will be given in the appendix.

About the year 1875 paragraphs were floating in the American newspapers concerning the " California flying machine," which was said to be under construction in San Francisco. This was the design of Mr. *Marriott*, of the San Francisco *Newsletter*, formerly a fellow-worker with Mr. *Stringfellow* in aeronautical pursuits, and also a resident of Chard.

Mr. *Marriott* had already experimented in 1867-69 with an elongated balloon, provided with attached aeroplanes, from which he expected to obtain additional sustaining power when at speed—a system which has been many times proposed, and which is often brought forward again by inventors who are not aware of the prior experiments.

Mr. *Marriott's* balloon model was 28 ft. long, 9½ ft. in diameter, with aeroplanes extending for half its length, and was to be driven by a light steam-engine, rotating a

propeller 4 ft. in diameter 120 times a minute.   The utmost
speed that could be obtained was five miles per hour, and
as this was not sufficient to stem the winds that constantly
prevail in San Francisco, the inventor turned his attention
to a design for an aeroplane.

This is said, in the report of the Aeronautical Society
of Great Britain for 1875, to have consisted of three planes,
superposed longitudinally, with an interval between them
of about 10 ft.   In transverse length the whole structure
was to be 120 ft. fixed upon a foundation of trussed bamboo, the planes being unequal in size, the largest on the
top being of the above dimensions and about 40 ft. wide,
the three planes being rigidly supported by two masts
about 40 ft. high and stayed by wire rigging.

To the lower end of each mast were to be affixed small
wheels, to run down an inclined rail, in order to impart
the necessary initial velocity to the apparatus, and this impulse was then to be continued by means of a steam-engine enclosed in a square compartment capable of holding
the engineers.   This compartment was to be located in
the center of the trussed bamboo keel.   The engine was
to work four screw propellers, two of them vertical and
two horizontal, their place of working breaking up the
continuity of the longitudinal planes.   The weight of the
whole machine was estimated at 1,500 lbs., including the
motive power and the engineer.

No drawings or detailed description of this aeroplane
were ever published, the inventor's idea being to keep his
plans secret until he had made a success of the machine.
It was never completed, for Mr. *Marriott* sickened and
died before the apparatus was ready for trial, and his associates did not care to risk the great outlay which would
have been necessary to test so large and expensive an apparatus.   The weight of the motor and the equilibrium
would have been the stumbling-blocks.

At the meeting of the Aeronautical Society of Great
Britain for 1876, M. *Sénécal* gave some notes on the stability of aeroplanes of different forms, which he illustrated
with paper models, and these experiments are so easily
reproduced that the following account of them, quoted
from the report, will probably prove interesting :

He said that while planes of even width and thickness (load
uniformly distributed) revolve upon their own axes, and their
path of translation is rectilinear, the motions of triangular
planes are much more complicated.   These planes are obtained
by dividing the circumference into blades of different widths.
These blades, besides revolving upon their axis, rotate also
round a vertical conic axis, whose base is upward, the vertex of
the plane describing a spiral round the conical axis.

He found that the rate of revolution and rotation increases in direct proportion as the base and the length of the blade decreases, and the length traveled over in a unit of time decreases also in the same proportion.   The shifting of the center of gravity (pressure ?) of these blades is most interesting.   It was found that the center of gravity of narrow planes was near the vertex and on the edge of the plane, but recedes toward the base and axis as it widens ; it also travels from the axis toward the edge and vertex as the rate of revolution increases, and possibly that, at high velocities of rotation, the center of gravity will be beyond the edge.   The size of blade that revolves and rotates most steadily represents the eighteenth to the twenty-fourth part of the circumference.   He also proved that by cutting a small plane out of the base it had the same effect as applying a weight at that point before cutting it.   The plane will then revolve and rotate round with its base turned toward the vertical axis.

He also liberated several narrow strips of paper, showing, while revolving, nodal and ventral sections similar to musical strings in vibration, the number of aliquot parts increasing with the length of the ribbons and disappearing as the width increases.

M *Sénécal* then enunciated the following law : that planes, of whatever form, but of even thickness and rigid margin, in order to translate steadily must carry their maximum load on a line representing the first third part from the anterior margins of the plane ; but one can, with impunity, apply graduated weights from that line right on to the edge, and, in some instances, a good distance beyond the edge, and high rate of speed is the result.   The rate of translation increases directly with the load placed on the different points of the graduations from that line of the center of gravity.

While the account of the action of these paper planes is not very clear, it is sufficiently so to permit the curious in such matters to repeat the experiments, and these will be found more instructive than any description of the results, however accurately expressed.   The action will be found to be greatly modified by slightly folding the back edges as already described.

In 1877 Mr. *Barnett*, of Keokuk, Ia., patented in the United States a flying machine somewhat similar in arrangement and principle with that of *Pénaud* and *Gauchot*. It consisted in an aeroplane something like a boy's triangular kite, but with the two longitudinal halves set at a diedral angle from the central spine or spar, in order to obtain lateral stability.   Just under this kite a boat-shaped car was to be affixed, carrying the motor, which was to rotate two propellers mounted upon shafts at the front of the apparatus, and turning in opposite directions.   An adjustable tail was to carry part of the weight and to regulate the angle of incidence, the car being provided with wheels so as to run over the ground until the speed was great enough to give a sustaining reaction.

This design is not without merit, but it leaves unsolved the two principal problems concerning aeroplanes—*i.e.*, the providing a light motive power, which shall not weigh more in proportion than that of the birds in ordinary flight, say 20 lbs per horse power, and the providing for automatic stability, which, as already explained, should be greater for an inanimate machine than for a live bird. The form of the triangular kite is not stable, as many a boy has found out to his sorrow by providing an insufficient tail, and if the kite form is to be used, it will probably be best to experiment with shapes that fly without a tail, some of which will be noticed hereafter.

Mr. *Barnett* is understood to have tried many experiments, extending over a period of 30 years. He first constructed a plain flat kite some 12 ft. long and 10 ft. wide, under which was hung a frame so as to attach and adjust the mechanism for turning two propellers rotating in opposite directions. This machine was not placed on wheels, but he was much pleased with the clutch that the propellers took on the air. Next he constructed an apparatus to carry the weight of a man. This consisted of a kite or aeroplane of canvas 27 ft. square, from which hung a propelling arrangement somewhat similar to that shown by Mr. *Maxim* in the *Century Magazine* for October, 1891, as the manner of connecting the aeroplanes and attaching the screws in his experimental apparatus. This machine was placed on wheels, being the running gear of a light spring wagon, and as Mr. *Barnett* knew of no motor sufficiently light for actual flight, he determined first to experiment with his own muscular power.

The propellers were two bladed, each blade being of oil-cloth and a sector of a circle, or like a piece from the ordinary round pie. The operator was beneath and rotated them through appropriate gearing. He ran the machine along smooth country roads, but as soon as speed was gained, the increasing air pressure, acting forward of the center of figure, in accordance with the law of *Joëssel*, already given, would tip up the front of the aeroplane and disturb the equilibrium. This led the inventor to believe that the propellers were too far below the aeroplane, and he altered their position, but without any better result ; the machine would still tip backward, presenting a greater angle of incidence, and increasing the resistance. Moreover, it would not keep to the line of the road, but, as it was propelled, would run off to either side into the grass, weeds or uneven ground, swerving in a way which would have involved great danger if it had been able to rise into the air.

Picking out a quiet evening, near dusk, the inventor determined to give it an extra good test over smooth ground, and while apprehensive that if it left the earth it might lurch and come to grief, he managed by " main strength and awkwardness" to get under considerable headway, when the front end tipped up so much as to break and splinter the main support, and the inventor came very near getting hurt.

This terminated the experiments with that machine. Subsequently the inventor entered it for exhibition at the State fair as an " automatic kite," and he says quaintly that he entered, at the same time, some samples of tomatoes, cabbage and grapes of his own growing ; received a premium on the tomatoes and cabbage, and favorable mention on the grapes, but concluded at the last moment not to take the " kite" to the fair ground, as it did not perform as he desired.

He has built two more machines of such dimensions as to support a man within the last six or seven years, and has tested them upon a smooth pasture, but found this, after many weeks of trial, too rough and uneven for his purpose. The tracks of animals, a bunch of grass, or a corncob would check the speed, so that with all his strength he could not arrive at sufficient velocity to leave the ground. This is not surprising, for Professor *Langley* has since shown that the best that can be done with a plane is to sustain 209 lbs. by the exertion of one horse power, and this without any hull resistance whatever, so that, as man cannot steadily exert much more than one-tenth of this power, a total weight of about 20 lbs. is the maximum that he can hope to support and drive through the air by the exertion of his own unaided strength.

Mr. *Barnett* has of course experimented with a considerable number of small models. He first tried clock springs, but found them too heavy ; and all would-be inventors had better avoid wasting effort with them ; next he tried twisted india-rubber, and while he found great irregularity in its action, he succeeded in obtaining a number of fair flights among many failures. He experimented with superposed planes, but the result was not satisfactory. His last model, produced in 1892, resembles his original design, and, driven by rubber bands, succeeded in getting a preliminary start by running over a platform 12 ft. long, slightly inclined, and flying through the air " above the hollyhocks and other flowers" until it struck the side of a house 30 ft. away, and 4 ft. higher than the platform.

India-rubber is a good reservoir of power to experiment with. The flights are brief, as the power is soon spent, but they give an opportunity of testing the equilibrium, the proportions, and the adjustment of the parts, which may suggest themselves to an experimenter as possibly efficient.

An apparatus patented in France by M. *Pomès*, in 1878, is represented in fig. 57. It consisted in two supporting planes in front, together with a keel plane, and a large

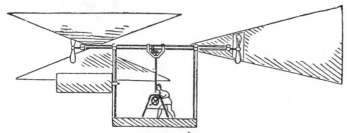

FIG. 57.—POMÈS—1878.

vane behind, to maintain the course. Two propelling screws on the same horizontal shaft were to impart motion, and although they are shown as actuated by hand on the figure, the same inventor had already patented, in 1871, in connection with M. *de la Pauze*, a gunpowder motor, in which a series of charges, exploded by electricity, were made to pass through a tube and to impinge against the buckets of a revolving wheel, from which the motion was to be communicated to the propellers. Neither this motor nor the aeroplane possess merit, and indeed the latter is about as badly arranged as it can be, for as the air pressures which are to sustain the weight act with a leverage increasing toward the tips of the wings, or sustaining planes, the latter should taper in plan from the center of the apparatus outward, instead of tapering inward as shown in the figure, in order to obtain a light and strong construction. It is not known whether M. *Pomès* experimented at all, but if he did, it must have been with very small models, for his design is quite unsuitable for a large one, and has been here included in order to point out the deficiencies of such a design.

In 1878 Mr. *Linfield* constructed an apparatus to test *his* conception of an aeroplane. It consisted of plane surfaces extended on a framework 40 ft. × 18 ft. at its greatest width, and measuring about 300 sq. ft. in surface, the weight of the apparatus being 189 lbs. It was mounted upon wheels, and driven over a macadamized road by the action of a screw propeller placed in front of the machine, rotated at about 75 revolutions per minute by the aviator, working a treadle and levers with cross handles. Upon the highway, on an incline of about one in a hundred, a speed of 12 miles an hour was attained without any indication of a rise from the ground. Then by going down hill, a speed of 20 miles per hour was obtained, but still without perceptible effect, which is not be wondered at, for at this speed, with an angle of incidence presumed to have been 6°, the "lift" would be 300 × 2 × 0.206, or, say, 123 lbs., while the weight including the aviator was over 300 lbs. It would have required an angle of 17°, at a speed of 20 miles per hour, to have produced sufficient "lift," while at that angle the "drift" alone would have required the exertion of 5 horse power, which the operator was clearly unable to furnish, it being "most dreadful exertion" to work the treadles at the flatter angle of incidence above presumed to have been experimented.

Subsequently Mr. *Linfield* built another machine upon a different principle. It was 20 ft. 9 in. in length, 15 ft. in width, and 8 ft. 3 in. high ; the sustaining surfaces being in two frames, each 5 ft. square. Each frame contained 25 superposed planes of strained and varnished linen 18 in. wide and spaced 2 in. apart, thus somewhat resembling a cupboard without front or back, and with shelves very close together. These frames were slung on either side of a cigar-shaped car at its maximum section, being set at a diedral angle to each other, so that the apparatus, could it have been seen in the air, would have resembled a huge cigar with a pair of saddle bags attached thereto. There was a nine-bladed screw at the front, and a guiding vane, like the tail of a dart, behind ; the entire sustaining surfaces in the two frames being estimated to aggregate 438 sq. ft., and the whole machine, which was mounted on four wheels, weighing 240 lbs., to which 180 lbs. must be added for the operator, thus providing a little over one square foot of sustaining surface per pound.

Mr. *Linfield* was to stand between the two front wheels and actuate two treadles to rotate the screw, which was 7 ft. in diameter ; but when the time arrived for testing the machine upon an ordinary macadamized road, it was stated that this could not be done on account of the impossibility of blocking the road during the trial. This was in a measure fortunate, for it led Mr. *Linfield* then to arrange with a railway to mount the machine on a flat car and to tow it behind a locomotive. When a speed of 40 miles per hour was attained the machine rose entirely free from the car, but was not allowed to swerve very far, as there was a side wind blowing, and it swung very close to the telegraph poles as it was. The tow line was some 15 ft. long, and the pull thereon was 24 lbs., which for a 240-lb. machine (without the aviator) indicates an angle of incidence of 1 in 10, or 6°. At this angle, and at a speed of 40 miles per hour, at which the air pressure would be 8 lbs. to the square foot, the total lift for a single plane ought to be $438 \times 8 \times 0.206 = 722$ lbs., so that, if the 240 lbs. of machine was just sustained, it indicates that the very narrow spacing (2 in.) between the superposed aeroplanes greatly interfered with their efficiency.

Mr. *Linfield* also tested the efficiency of superposed screws. He placed nine of them some 6 in. apart upon a vertical shaft. These were all with two narrow blades and 3 ft. in diameter, but in whatever relative position they were placed radially, he could get no greater lift from the nine screws than he could from the top and bottom screws only, 4 ft. apart, the seven intermediate screws being removed.

The idea of testing the apparatus by towing it on a railway car was evidently a good one, but this disclosed such inefficiency of lifting power and of stability as to put an end to the experiments.

We next come to a series of very careful experiments, tried by an able mechanician, which almost demonstrate that artificial flight is accessible to man, with motors that have been developed within the last two years. These experiments were carried on by M. *V. Tatin*, who was then Professor *Marey's* mechanical assistant. He first began with beating wings, and produced, in 1876, the artificial bird which has already been briefly noticed under the head of "Wings and Parachutes." This was driven by twisted rubber ; not only did M. *Tatin* find that the power required was unduly great, but he also found that this power could not be accurately measured, the torsion of india-rubber being erratic and stretching unequally. He constructed a large number of mechanical birds of all sizes and various weights ; he tried many modifications and entire or partial reconstructions, and finally concluded, after spending a good deal of time and money, to take up the aeroplane type, to be driven by a reservoir of compressed air. With this his efforts were successful almost from the first, and he produced in 1879 the apparatus shown in fig. 58, which is practically the first that has risen into the air by a preliminary run over the ground. This machine consisted in a silk aeroplane, measuring 7.53 sq. ft. in surface, being 6.23 ft. across and 1.31 ft. wide, mounted in two halves at a very slight diedral angle, on top of a steel tube with conical ends which contained the compressed air. This reservoir was $4\frac{3}{4}$ in. in diameter and $33\frac{1}{2}$ in. long, was tested to a pressure of 20 atmospheres, and worked generally at 7 atmospheres ; its weight was only 1.54 lbs., and its cubical capacity 0.28 cub. ft. From this (the vital feature of the machine) the stored energy was utilized by a small engine, with oscillating cylinder, placed on a thin board on top of the tube, and connected by shafts and gearing to two propellers with four vanes each, located at the front of the aeroplane. These propellers were 1.31 ft. in diameter, and rotated in opposite directions some 25 turns per second, their velocity at the outer end being about 100 ft. per second. The vanes were of thin bent horn set at a pitch of about 1.50 ft., and they towed the apparatus forward instead of pushing it.

A tail of silk fabric 1.97 ft. across at the rear, by a length of 1.97 ft., was set at a slight upward angle and

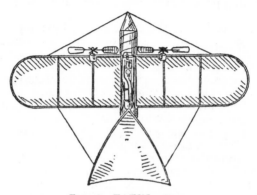

FIG. 58.—TATIN—1879.

braced by wire stays, in order to provide for the longitudinal stability upon the principle advanced by *Pénaud ;* and the whole apparatus was placed on a light running gear consisting first of four wheels, and subsequently of three wheels. The total weight was 3.85 lbs., so that the sustaining surface of the aeroplane (omitting the tail) was at the rate of 1.95 sq. ft. to the pound.

After a vast deal of preliminary testing and adjustment, the apparatus was taken to the French military establishment at Chalais-Meudon, where it was experimented with in 1879 upon a round board platform 46 ft. in diameter. Upon this the machine would be set upon its wheels, the front and rear ends being fastened to two light cords carried to a ring around a central stake, and the compressed air would be turned on to the engine. The propellers would put the apparatus in motion, and it would run from 65 to 165 ft. over the boards, until it attained a velocity of 18 miles per hour, when it would rise into the air, still confined radially by the two cords, and make a flight of about 50 ft., when, the power being exhausted, it would fall to the ground, almost invariably injuring the running gear in doing so.

The flights were not very high, but on one occasion the apparatus passed over the head of a spectator. The angle of incidence was 7.° or 8°, and the power developed by the engine was at the rate of 72 33 foot-pounds per second, gross ; but as its efficiency was only 25 to 30 per cent. of the gross power, the effective force was at the rate of 18.08 to 21.70 foot-pounds per second, or, say, at the rate of 5 foot-pounds per second (300 foot-pounds per minute) per pound of apparatus.

This power was measured with great care, the machine being provided with a tiny gauge and tested repeatedly with a dynamometer. M. *Tatin* calls attention to the fact that the minuteness of the engine greatly diminished its efficiency, and that with large machines it would be comparatively easy to obtain 85 per cent. of the gross power developed. He draws the conclusion that his apparatus demonstrates that 110 lbs. can be sustained and driven through the air by the exertion of 1 horse power—a most important conclusion, which will be further discussed hereafter.

To return, however, to the experiments : they are described as follows by M. *Tatin.**

I will pass without description a series of preliminary experiments which led me to modify certain details, until all conditions were favorable. I then had the satisfaction of seeing the apparatus start at increasing speed, and in a few seconds the carriage barely touches the ground ; then it leaves it entirely at a speed of about 18 miles per hour, which agrees closely with the calculations. It describes over the ground a curve similar to those described by small models gliding freely, and when it comes down after its orbit, the shock is so violent as to injure the running gear. This accident recurred upon each experiment carried out under the same conditions : the carriage was soon destroyed, and even the propellers were injured, although they could be repaired. I then tried another experiment, which I had already attempted several times without success, in consequence of inadequate preparation. The apparatus, the running gear being removed, was suspended by two grooved wheels running freely over an iron telegraph wire 260 ft. long, stretched as rigidly as practicable. When the speed became sufficient, the apparatus rose, and then one of the propellers struck the iron wire ; the front grooved wheel overtook the machine, and the propeller was destroyed. These accidents caused no repining, for they demonstrated that in all cases the apparatus had completely overcome the force of gravity.

In order to continue the experiments I built a new carriage

---

* *Aéronaute*, September, 1880.

and new propellers, hoping to make them strong enough to stand the shocks during a new set of experiments, from which to deduce accurately the work done. The new running gear had but three wheels, these being larger and lighter than the old. The propellers, on the other hand, were made heavier, but modified so as to rotate more easily. Their vanes were made of a thin sheet of horn bent hot to the proper curvature. The inner two-fifths from the hub consisted of steel wire, this portion of a propeller requiring much force for rotation, and giving out but small effect toward propulsion ; but the diameter and the pitch were the same as formerly.

I was, unfortunately, unable to make all the experiments I desired with this repaired apparatus. I intended to study the results with various angles of incidence in the planes and various pitch of the propellers ; then to study the important question as to the best proportion between the sustaining surface and the diameter of the propellers ; and lastly the speed of translation which will best utilize the force expended.

I was nevertheless enabled to deduce the following figures from my experiments. These figures are not absolutely exact, but sufficiently so to serve as a guide to others who may wish to engage in similar work. Calling A the sustaining surface in square meters (without the tail). and V the speed of translation in meters per second. then we may say :

$$\text{Lift} = O. \; kg. \; .045 \; A \; V^2.$$

And the motor will need to develop effective work at the rate of 1.50 kilogrammeters per kilogramme of the weight (4.935 foot-pounds per second per pound), which corresponds to one horse power for each 110 lbs. weight of the apparatus.

These experiments seem to demonstrate that there is no impossibility in the construction of large apparatus for aviation, and that perhaps even now such machines could be practically used in aerial navigation.

Such practical experiments being necessarily very costly, I must, to my great regret, forego their undertaking, and I shall be satisfied if my own labors shall induce others to take up such an enterprise.

The effective work done by this aeroplane having been accurately measured, it affords a good opportunity of testing the method of estimating resistances which has been proposed by the writer in estimating the work done by a pigeon.

The weight of M. *Tatin's* apparatus was 3.85 lbs. Its aeroplane surface was 7.53 sq. ft., the angle of incidence was 8°, and the speed was 18 miles per hour, at which the air pressure would be 1.62 lbs. per sq. ft. Hence we have, by the table of " lift and drift " :

$$\text{Lift, } 8° = 7.53 \times 1.62 \times 0.27 = 3.29 \text{ lbs.,}$$

which indicates that a small part of the weight was sustained by the tail.

The hull resistances are stated by M. *Tatin* to have been almost equal to that of the plane. These hull resistances would consist of that of the tube, of 0.12 sq. ft. midsection, which, having conical ends and parallel sides, will have a coefficient of about one-third of that of its midsection. The resistance of the wheels and running gear will be slightly greater, but must be guessed at, as the wheels would continue to revolve through inertia and thus increase the resistance.

The front edge of the aeroplane, which was of split reed and about one-eighth of an inch thick, was 6.23 ft. long ; but as the back edge of the aeroplane and the side borders of the tail would also produce some air resistance, we may call the edge resistance as equal to 6 ft. in length, by a thickness of 0.01 ft., without any coefficient for roundness. We then have the following estimate of resistances :

RESISTANCE OF TATIN AEROPLANE.

Drift 8°   — 7.53 × 1.62 × 0.0381 = 0.4648 lbs.
Tube   — 0.12 × 1.62 ÷ 3   = 0 0648 "
Wheels and gear — estimated   = 0.1000 "
Edges of wings — 6 × 0.01 × 1.62   = 0.0972 "

Total resistance = 0.7268 "

and as the speed was 18 miles per hour, or 26.40 ft. per second, we have for the effective power required :

Power = 0.7268 × 26.4 = 19.19 foot-pounds per second,

which agrees very closely with the 18.08 to 21.70 foot-pounds per second said to have been effectively developed, and is at the rate of 5 foot-pounds per pound of apparatus, or of 110 lbs. of weight per horse power.

This last is the important point. Now that Mr. *Maxim* has produced a steam-engine which, with its boilers, pumps, generators, condensers, and the weight of water in the complete circulation, weighs less than 10 lbs. to the horse power, aviation seems to be practically possible, if only the stability can be secured, and an adequate method of alighting be devised.

Ocular demonstration being always more satisfactory than description, those readers who have been sufficiently interested in the subject to try the experiments which have been described with paper planes (falling by gravity) may also like to see for themselves how an aeroplane behaves when motive power is applied. They can probably obtain in a shop one of the toys which have already been alluded to, under the head of "Screws to Lift and Propel," as one of the series produced in 1879 by M. *Dandrieux*, and which is shown in fig. 59.

This is a true aeroplane, the wings being fixed, and the propulsion being produced by the screw at the front, which represents the antennæ of the butterfly. This screw is driven by the unwinding of the rubber threads, and has practically no pitch except that produced by the yielding of the posterior edge of the gold-beater's skin, of which the vanes are composed. Its peculiar shape, giving a

FIG. 59.—DANDRIEUX—1879.

maximum of surface near the outer end, with a rigid anterior edge and an elastic posterior edge, is the result of a good deal of experiment, and may furnish a useful hint for those desiring to experiment upon a larger scale. The wings are also of gold-beater's skin, and instead of being stretched tightly upon the frame, the anterior margin only is made rigid, the rest of the surface being left quite loose, so that it may undulate when under forward motion, as in the case of M. *Brearey's* device, which will presently be described. This feature in construction, which differs greatly from that which obtains in the case of birds and insects, whose wings are elastic, but do not undulate, is said to be intended to compensate for defects in workmanship and equilibrium. Upon being tested in still air within doors, the toy will be found quite erratic in flight. It will generally go up to the ceiling, and then flutter in various directions until the power is exhausted, and seldom twice pursue the same course. Out-of-doors it will rise some 20 or 30 ft., dart about, or drift with the wind, until the rubber threads are unwound, and then glide down to the ground sustained by its aeroplane alone.

As a matter of course the sustaining surfaces have to be made very large in proportion to the weight, in order to prevent injury in alighting. One of these little toys, computed by the writer, weighs 86 grains or 0.0123 lbs., and measures 50 sq. in. in aeroplane surface, or 0.3472 sq. ft. ; this being in the proportion of 28 sq. ft. to the pound, or about 0.7 of that of the real butterfly, which, being much smaller, measures some 40 sq. ft. to the pound, and which

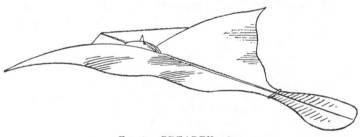

FIG. 60.—BREAREY—1879.

in consequence is capable of but slow flight, although it is not infrequently found by aeronauts floating about in the upper air a mile or so above the earth, a fact to which further reference will be made when we come to consider the prevalence of upward trends in aerial currents.

The propulsion of a loose undulating surface was at about the same time, somewhat differently and quite independently, proposed by M. *F. W. Brearey*, the Honorary Secretary of the Aeronautical Society of Great Britain. He patented, in 1879, the apparatus shown in fig. 60, in which a flexible fabric is attached to a central spine and to vibrating wing arms at the front, which latter beat up and down like the wings of a bird. The effect of this action is to throw the fabric into a state of wavelike motions, both lengthwise and in a smaller degree also laterally, which are said to cause the apparatus to be both supported and propelled in the air, while an adjustable tail regulates the angle of incidence. The wing arms are flexible and stayed to a bowsprit by cords, and the power for an actual machine is to be placed in a car or body affixed along the central spine.

M. *Brearey* records that he took the idea from watching the movements of a " skate" fish in an aquarium, which in swimming undulated its whole body, and that he found that when applied to propulsion in air the loose fabric greatly added to the stability, so that the device might be considered as a sort of dirigible parachute, which would come down safely if the motive power became exhausted from any cause. In the various models which he made to illustrate the experimental lectures, with which he was accustomed to popularize " the problem of flight" in Great Britain, he used the torsion of india-rubber to produce the revolution of the crank which vibrated the arms, thus getting a dozen strokes or so, and he claimed that the smaller model (5 ft. $\times$ 8 ft.) flew from his hand, on one occasion at least, perfectly horizontally to the extent of 60 ft., no angle of incidence of the apparatus being perceptible. The larger model was 6 ft. wide by 10 ft. long, with about 16 sq. ft. of surface, and a weight of 3.1 lbs. (of which o 44 lbs. was added ballast, which it easily carried), being thus in the proportion of some 5.15 sq. ft. per pound of weight, with which the falling velocity would be about 9 ft. per second, or equal to a descent from a height of 1.27 ft., but which was nevertheless found to be too heavy to be safely used in public experiments over the heads of an audience. From his experiments M. *Brearey* drew the following conclusions as to the possibilities of his apparatus :

We are thus at liberty to contemplate the construction of an aerial vehicle whose dimensions would suffice to maintain, in wave-action, 600 or 700 sq. ft. of canvas, actuated by steampower, and capable of supporting the additional weight of a man, whose weight, together with the machine, would certainly not exceed 500 lbs. ; and we can contemplate the man as being able to move a few feet backward or forward without much affecting the stability of the machine. His descent under the parachute action can thus be graduated at will. This can also be effected by a cord attached to the tail, which by that means can be elevated or depressed at pleasure. Placed upon wheels it has, of course, yet to be ascertained what distance of preliminary run would be required, assisted by the action of the fabric, before it would rise from the ground.

Subsequently (his second American patent is dated in 1885) M. *Brearey* further proposed the superposition of two or more sets of such " wave-action" aeroplanes, and the important addition of what he calls the " pectoral cord," which consists in an elastic cord (or suitable spring) attached to some point underneath each of the lower set of wing-arms and passing underneath the carriage, car or central spine, so that it may be thrown into tension on the up stroke, and restore the power thus stored upon the down stroke of the wing-arms. This device is designed to imitate in its action the functions of the pectoral muscle of a bird. The tension of this cord or spring is regulated in accordance with the weight to be sustained, and is said to be perfect when, upon the whole apparatus being committed free to the air, the wing-arms are retained at a suitable diedral angle against the upward pressure. It follows from this action that the up stroke, being assisted by the air pressure which sustains the weight of the apparatus, expends less power than the down stroke, and that nearly all the power can be used in depressing the wing-arms to compress a wave of air, which undulating backward and outward along the loose fabric may assist the air pressure already due to the forward speed, in sustaining the aeroplane, and serve at the same time to propel it.

M. *Brearey*, however, seems to have applied this " pectoral cord" chiefly to those of his models which showed the wing-action proper, and in the practical demonstration which he gave to the Aeronautical Society of Great Britain, at its meeting in 1882, he said :

Working in the field of experiment, I am enabled to state that the power requisite to propel and sustain a body in the air has been greatly overestimated, even by those who took the more favorable estimate in view of the ultimate attainment of flight. I am not aware, however, that the true reason for the minimum display of actual power exerted in the flight of birds has ever been propounded. Certainly it has never before been demonstrated by actual experiment.

The action of the pectoral muscles of the bird alone accounts for this. Consequently the advantage would be altogether lost in anything but a reciprocal action. The bird commits himself to the air, and the pressure of the air underneath the wings forces them upward. The weight of the bird is indicative of the pressure ; and as a consequence of this automatic raising of the wing by the pressure of the air underneath, we should imagine that the elevator muscle need not be strong. As a matter of fact, we find it is weak. I doubt whether any muscular effort is made to elevate the wing at all in flight ; but when not in flight, the bird of course requires the power, to elevate its wing in preparation for it.

Committed, then, to the air, the elastic ligaments connected with the wings are stretched to that degree which allows of the wings being sufficiently raised for effective support without flapping, and without, as I conceive, any muscular exertion upon the part of the bird. The limited power of the elevator muscle may here come into use occasionally in aid of the under air-pressure, and with the further effect of stretching the ligaments. Now it will be argued that in the downward stroke there must be as much muscular force employed as will raise or, at least, prevent from falling, the weight of the bird ; but this is not so, because the reaction of these ligaments, which have been stretched entirely by the weight of the bird, assists materially the action of the depressor muscle. . . . .

M. *Brearey* here produced a model having wings measuring 4 ft. from tip to tip. He showed the elastic cord underneath the wings, but for the purpose of the first experiment he detached it. He then wound up the india-rubber strands 32 times, and showed that this, although sufficient to flap the wings with energy while held in the hand, was insufficient to cause the model to fly. This was demonstrated by letting the model free. He explained its inability to fly from its want of power to bring the wings down with sufficient force.

He now unwound the action and proceeded to wind it up again 32 times, and attached the pectoral cord. Holding the model in his hand, he called attention to the fact that it was powerless to flap the wings because the two forces were in equilibrium. It required the addition of another force to effect flight, and he asked what that other force could be except weight ? If now it flew, he proved beyond the possibility of doubt that weight was a necessity for flight. The model was then set free, and flight was accomplished.

He also showed that the model would only fly without the attached pectoral cord when wound up 40 times. With the cord it would fly when wound up only 13 times, thus showing the great saving in power which accrued through the action of the pectoral cord.

M. *Brearey* then produced a model of his " wave aerial machine," having 4 sq. ft. of loose surface weighted to ½ lb., and he demonstrated by its flight that the principle was equally applicable to that.

It may be questioned whether this " wave action" is likely to prove economical of power in either sustaining or propelling an aeroplane, for it seems difficult to conceive that a wave of air compressed at the front by the wing-arms should travel back to the rear, unconfined as it is either at the bottom or sides. Still, the loose surface may add to the stability, as claimed for the *Dandrieux* toy, and it would certainly diminish by its yielding the strains that would otherwise occur at the points of attachment of a rigid surface in an aeroplane ; but M. *Brearey's* wave-action seems to be chiefly applicable as a dirigible parachute, and a small model upon this principle, but without motive power, was once liberated as an experiment by Captain *Templer*, from a balloon which had risen 200 ft.

or 300 ft. from Woolwich Arsenal, and it traveled back again to the arsenal, half a mile, against the wind.

It seems somewhat singular that so few efforts have been made to devise dirigible parachutes, a system which M. *de la Landelle* constantly extolled, as constituting the first requisite step toward eventual flight by working out the problem of absolute stability and safety. The only one of these devices which the writer has been able to find recorded is that of M. *Couturier*, patented in France in 1875, and this is so briefly described in the *Aéronaute* for November, 1878, that its mode of operation cannot be made out.

The " pectoral cord" attachment is probably a valuable device for flapping wings, as furnishing that inequality of effort between the up and the down stroke which undoubtedly obtains in bird flight. This effect was produced in a " wave-action" model exhibited by M. *Brearey* at the aeronautical exhibition of the Aeronautical Society of Great Britain of 1885, by a " trunk engine" designed and built by M. *Hollands*, which, however, was not shown under steam, as the boiler was only just completed in time for the exhibition ; but M. *Hollands* said that the model flew well, and supported weights, when the engine was supplied with compressed air through an india-rubber tube. He does not seem to have stated what power was exerted.

While almost all inventors and experimenters of aeroplanes have proposed some sort of motive power, and have found their designs paralyzed very soon by the want of a sufficiently light motor, there have been at various times, as already intimated, keen observers of the flight of soaring birds, who have held that once under way in a sufficient breeze, the performance involves no muscular movement whatever, save in balancing, and that the wind alone furnishes sufficient motive power (if blowing from 10 to 30 miles per hour) to enable man to soar and to translate himself at will in any direction, even (paradoxical as it may seem) against the wind itself.

Chief among these observers in recent days stands M. *Mouillard*, of Cairo, Egypt, who has spent over 30 years in watching birds soar in tropical latitudes, and who published, in 1881, a very remarkable book (in French), " L'Empire de l'Air," which should be read by all those seriously interested in the solution of the problem of flight. This book, the result, as the author explains, of a passionate vocation which began at the age of 15, is almost wholly a record of personal observations and deductions. Its sub-title designates it as an " essay upon ornithology as relating to flight," but it is far more than that, for it not only describes the flight and manœuvres of birds, and gives good reasons for the author's belief that they can be imitated by man, but it describes four attempts which he has made to do so with various forms of apparatus.

M. *Mouillard* underrates, perhaps, the value of mathematical investigation, and he sometimes errs in his explanation of physical phenomena ; but his observations are unrivaled, and they are presented with a particularity of circumstance, a vivacity and a charm which photograph them at once on the mind of the reader. He begins by explaining the difference between useful and unfruitful observations of creatures so willful, so swift, and so shy as the birds ; then he describes the various modes of flight (both rowing and sailing), and the movements of the various organs, such as the wings and the tail ; the influence of their shape in determining the mode of progression and the speed of the various species, and he shows conclusively that if these organs are properly shaped therefor, the heavier the bird the more perfectly he soars, and can, once initial speed is gained, *sail indefinitely upon the wind without further flapping his wings*. This is the keynote of the book ; observation after observation is described, anecdote after anecdote is related, to impress upon the reader that there need be no flapping whatever, if only the wind be strong enough ; and that when there is no wind, the soaring bird must come down to the ground or resort to flapping, like the rowing birds.

Then the effect of the speed of the wind is discussed. It is shown that certain species of soaring birds with broad wings, such as the kites, the eagles, and the vultures can sail upon a wind blowing at 10 to 25 miles per hour, but must seek shelter when it increases to a gale, while the sea-birds, with long and narrow wings, such as the gulls, the frigate bird, the albatross, sport indefinitely in the tempest blowing at 50 or more miles per hour. He arrives at the conclusion that when man succeeds in imitating the manœuvres of the soaring birds, he will utilize the moderate winds, and attain to speeds of about 25 to 37 miles per hour.

M. *Mouillard* also passes in review the individual mode of flight and characteristics of the various species of birds, both the rowers and the sailers ; comprising some 13 different types, and giving tables from his own measurements of weights, surfaces, dimensions, etc., which have been compiled by M. *Drzewiecki*, and have already been quoted by the writer under the head of " Wings and Parachutes ;" while he finally expresses a strong opinion that the easiest type for man to imitate is the great tawny vulture of Africa (*Gyps fulvus*), which weighs some 16.50 lbs., and spreads some 11 sq. ft. of surface to the breeze.

M. *Mouillard* explains how, in his opinion, the manœuvres of this bird can be imitated, so as to obtain both a sustaining and a propelling effect from the wind, and he describes (much too briefly) the four several attempts which he had then made to demonstrate the correctness of his theory of the possible soaring flight of an aeroplane for man.

The third of these aeroplanes, as described in 1881, is shown in fig. 61. It consisted of two thin boards, properly stiffened, to which were attached ribs of "agave" wood (an African aloe, exceedingly light and strong), which ribs carried the fabric constituting the two wings. The two boards were hinged vertically together (somewhat imperfectly) at the center, and the operator stood upright in the central space at *c*, suspended by four straps attached to the boards near the hinge ; two of these straps passing over the shoulders and two between the legs. Moreover,

Fig. 61.—MOUILLARD—1865.

light wooden rods extended from the feet to the outer ends of the boards, so that the angle of the wings with each other could be varied at pleasure.

Standing upright, with this apparatus strapped on, the hinge was about at the height of the pit of the stomach, the arms being extended out flat upon the boards, and slipping under straps ; M. *Mouillard* trusting to such shifting of his body within the space *c* as he could effect by resting his weight on his arms, to produce the necessary changes in the center of gravity of the apparatus, which were required by the changes in the angles of incidence.

The whole apparatus weighed 33 lbs., but was found unduly light, as the parts yielded and the wood cracked when tested with vigorous thrusts of the legs. It had been hastily constructed, with such materials as the country afforded, and the builder was not satisfied with it.

M. *Mouillard* gives but a scanty description of his experiments with this aeroplane in "L'Empire de l'Air," so little, indeed, as to suggest further inquiry ; but he has since written another book, which he entitles "Le vol sans battements'' (flight without flapping), which is now nearly ready for the press, and wherein he records further observations, explains more fully his ideas and the results of his meditations, giving freely, as he expresses it, '' all that he knows,'' and in which there is a fuller account of the experiment in question.

From this forthcoming book M. *Mouillard* has kindly furnished the following extract concerning the experiment with the apparatus shown in fig. 61.

It was in my callow days, and on my farm in the plain of Mitidja, in Algeria, that I experimented with my apparatus, No. 3, the light, imperfect one, the one which I carried about like a feather.

I did not want to expose myself to possible ridicule, and I had succeeded by a series of profound combinations and pretexts in sending everybody away, so that I was left all alone on the farm. I had already tested approximately the working of my aeroplane by jumping down from the height of a few feet. I knew that it would carry my weight, but I was afraid to experiment in the wind before the home folks, and time dragged wearily with me until I knew just what the machine would do ; so I finally sent everybody away—to promenade themselves in various directions—and as soon as their backs were turned, I strolled into the prairie with my apparatus upon my shoulders. I ran against the air and studied its sustaining power, for it was almost a dead calm ; the wind had not yet risen, and I was waiting for it.

Near by there was a wagon road, raised some 5 ft. above the plain. It had thus been raised with the soil from ditches about 10 ft. wide, dug on either side.

Then came a little puff of wind, and it also came into my head to jump over that ditch.

I used to leap across easily without my apparatus, but I thought that I might try it armed with my aeroplane ; so I took a good run across the road, and jumped at the ditch as usual.

But, oh horrors ! once across the ditch my feet did not come down to earth ; I was gliding on the air and making vain efforts to land, for my aeroplane had set out on a cruise. I dangled only one foot from the soil, but, do what I would, I could not reach it, and I was skimming along without the power to stop.

At last my feet touched the earth, I fell forward on my hands, broke one of the wings, and all was over ; but goodness ! how frightened I had been ! I was saying to myself that if even a light wind gust occurred, it would toss me up 30 to 40 ft. into the air, and then surely upset me backward, so that I would fall on my back. This I knew perfectly, for I understood the defects of my machine. I was poor, and I had not been able to treat myself to a more complete aeroplane. All's well that ends well. I then measured the distance between my toe marks, and found it to be 138 ft.

Here is the *rationale* of the thing.   In making my jump I
acquired a speed of 11 to 14 miles per hour, and just as I
crossed the ditch I must have met a puff of the rising wind.   It
probably was traveling some 8 to 11 miles per hour, and the
two speeds added together produced enough pressure to carry
my weight.

I cannot say that on this occasion I appreciated the delights
of traveling in the air.   I was too much alarmed, and yet never
will I forget the strange sensations produced by this gliding.

Then M. *Mouillard* repaired the injured aeroplane, and
he tried it again a few days later.   Of this later experiment he says in " L'Empire de l'Air" :

I had no confidence, as I have already stated, in the strength
of my aeroplane.   A violent wind gust came ; it picked me up ;
I became alarmed, did not resist, and allowed myself to be upset.   I had one shoulder sprained by the pressure of the two
wings, which folded up against each other like those of a butterfly when at rest.

M. *Mouillard* then determined to make no more experiments with this incomplete machine, but to build a better
one, with which he could control all the manœuvres necessary for soaring, but shortly afterward his circumstances
led him to leave the farm and to remove from Algeria to
Cairo, Egypt.   Here, in a great city, he no longer had the
facilities for experimenting that he possessed on the farm,
for he had to go out some distance to secure space and
privacy for each experiment.   Then came illness ; the
former gymnast became a cripple, so that he could no
longer perform for himself the acrobatic manœuvres
necessary to experiment with a soaring apparatus, but
still he persevered, and he describes in " L'Empire de
l'Air" the design for the fourth apparatus, of which he
began the construction in 1878, but which was interrupted
by ill-health.

Since the publication of his book in 1881, M. *Mouillard*
is understood to have been continuously engaged in perfecting and simplifying his proposed soaring apparatus,
and in trying experiments (by proxy) with models on a
small scale.   He says that he will soon be prepared to
have the matter tested on a large scale, and that he has
never wavered from absolute conviction in the truth of the
principles which he laid down in " L'Empire de l'Air," in
which he expresses himself as follows :

I hold that in the flight of the soaring birds (the vultures, the
eagles, and other birds which fly without flapping) *ascension is
produced by the skillful use of the force of the wind, and the steering, in any direction, is the result of skillful manœuvres ; so that by
a moderate wind a man can, with an aeroplane, unprovided with*

any motor whatever, rise up into the air and direct himself at will, even against the wind itself.

*Man therefore can, with a rigid surface and a properly designed apparatus, repeat the exercises performed by the soaring birds in ascension and steering, and will need to expend no force whatever, save to perform the manœuvres required for steering.**

" The exact shape of these aeroplanes need not be discussed in this chapter, for it will be seen further on that there are scores of shapes and devices which can be employed, but all forms of apparatus, however dissimilar, must be based upon this idea, which I repeat."

Ascension is the result of the skillful use of the power of the wind, and no other force is required.

M. *Mouillard* then continues :

It will doubtless be very difficult for many persons to admit that a bird can with a moderate wind, remain a whole day in the air with no expenditure of power. They will endeavor to suppose some undetermined pressures or some unseen flappings. In point of fact, the human understanding does not readily admit the above truth ; it is astonished, and seeks for all the evasions it can find. All those who have not *seen* say, when ascension without expenditure of force is mentioned to them, " Oh, well, there were some motions which escaped your observation."

It even occurs sometimes that a chance or superficial observer, who has had the luck to see this manœuvre well performed by a bird, when he turns it over in his mind afterward feels a doubt invading his understanding ; the performance seems so astonishing, so much against ordinary experience, that the man asks himself *whether his eyes did not deceive him.* For this observation, in order to carry absolute conviction, must bear upon the performance of the largest vultures, and they alone ; and this is the reason : it is because all the other birds which ascend into the air by this process do not perform the necessary decomposition of forces required in all its naked simplicity.† . . . .

To be convinced, a man must *see ;* for to see the performance even once is better than a whole volume of explanations. Therefore, O reader, if you are interested in this subject, go and see for yourself, and be edified. Go to the regions where dwell the birds which perform these demonstrations ; and when you have beheld them for a few instants, being already initiated as to what to observe, comprehension will at once come into your understanding.

Whoever has seen a boy's kite ascend into the air, and considered that the string may be replaced by a weight, *if only the equilibrium be secured and maintained,* will

---

* The italics are M. *Mouillard's* own.
† The present writer has seen the feat performed by gulls many times.

have no difficulty in granting the correctness of M. *Mouillard's* assertion that the power of the wind is quite sufficient to secure ascension, but it will not so readily be understood how it is also sufficient to secure progression even against the wind. It will, indeed, be conceived that an aeroplane possessed of initial velocity can soar in a circle in the wind like a bird, and by changing its angle of incidence, descend somewhat when going with the wind, and rise again in consequence of the greater " lift" when facing the wind, thus gaining in height at every lap, and eventually utilizing the elevation thus gained in gliding in any desired direction, *always provided that the equilibrium be maintained ;* but this involves very delicate manœuvres, which will be further considered when we come to sum up the results of all the experiments with soaring devices, and indeed the subject warrants a paper by itself, which may be placed in an appendix.

It may, however, be said here that the French aviators, after having long doubted the reality of the performance of sailing flight by the birds, whose evolutions they were unable to watch in their climate, have had so many corroborations furnished to them by trustworthy witnesses, that they now generally admit that a soaring bird can sustain himself indefinitely on a wind, without flapping, and that man may learn to imitate him if only a proper apparatus be designed, and the operator possesses the necessary knowledge and skill to work it, so as to perform the right manœuvres and at the right time.

But these wonderful performances of the " sailing birds" are chiefly witnessed in tropical or semi-tropical regions— those in which the steady trade winds or the regularly incoming sea breezes afford daily to the birds the power of performing their evolutions in search of food. In the more temperate regions the wind does not blow every day with just the right intensity, the casual soaring bird is frequently compelled to resort to flapping, and the would-be inventor has his thoughts directed to some form of a power machine; generally some combination of aeroplanes with propelling screws, which differs a good deal from the simple, compact, and severe outlines indicated by nature.

The form of the soaring bird is reducible to three elements. First, a comparatively large body, shapely, but unsymmetrical fore and aft, presumably the solid of least resistance to the air ; second, a symmetrical pair of wings, convex on top, and more or less concave beneath, with a sinuous and stiff front edge ; and, third, a tail, which varies greatly in its proportion among the various species. For these features, most of the inventors have substituted a small, boat-like body, a combination of flat planes and flat rudders, both horizontal and vertical, which last is not found to exist in the bird.

A good case in point is found in the instance of Mr. *Krueger*, who patented in the United States, in 1882, a flying machine consisting in three flat horizontal planes set one behind the other, the front one being triangular in plan, while the rear one might be shaped like the tail of the swallow. These were to be adjustable, so as to guide the machine up or down. Beneath them was to hang a ship or vessel, and above them were to be set still other planes, sloping like the two sides of a roof, in order to act as a parachute. Four propelling screws were to be arranged between the three sustaining planes, while four adjustable keel cloths, vertically affixed both above and below the sustaining planes, were to steady the course and to furnish the steering power.

No particular motive power was proposed, and no method indicated for maintaining the stability, so that it is quite safe to say that no experiments were ever tried with this apparatus upon any practical scale. It has been here mentioned to illustrate how misguided ingenuity sometimes runs to complications, while leaving untouched the really essential requirements.

The next inventor to be noticed, M. *Goupil*, a distinguished French engineer, began otherwise : by taking thought of the motive power and of the equilibrium. After having tried a few preliminary experiments, he designed in detail a light steam-engine and boiler, the weight of which he estimated at 638 lbs. for a machine of 15 horse power gross, or 42.5 lbs. per horse power. He also designed a condenser of like capacity, estimated to weigh some 220 lbs. (15 lbs. per horse power), so that the water could be used over and over again ; and he then figured that the rest of the flying apparatus, without cargo, might weigh 242 lbs., thus making a total of 1,100 lbs., so that if the steam-engine worked up to two-thirds of its theoretical efficiency and developed 10 effective horse power, the total apparatus would have been in the proportion of 110 lbs. per horse power, but might be reduced to about 44 lbs. per horse power through the use of aluminium instead of other metals.

These estimates of weights of motor and condenser have been since then more than confirmed by the achievements of M. *Maxim* and other inventors, but before seeking to realize them M. *Goupil* determined to investigate the all-important question of equilibrium.

Both observation and mathematical considerations had satisfied him that much of the longitudinal stability of the bird in the air was due to the raking shape, fore and aft, of the under part of its body, which, presenting to the air an increasing and more effective angle of resistance when pitching oscillations occur, tended to restore the balance and to prevent the animal from taking either a " header" or a " cropper." This he determined to test experimentally, and he accordingly built, in 1883, an apparatus similar to that represented by fig. 62 ; omitting, however, the screw, the lower framework, and the stays to the wings.

The alar spread was 19.68 ft. from tip to tip of wings, the length was 26.24 ft. from the head tip to the end of the tail, and the mid-section was 26 90 sq. ft. in area, while the sustaining surface was no less than 290 sq. ft., the weight being 110 lbs.

It will be noticed that this was a marked departure from

FIG. 62.—GOUPIL—1883.

the ordinary aeroplane types, there being an ample body to contain machinery, and the wings being decidedly concavo-convex, while other inventors have generally endeavored to diminish the body as much as possible and to gain support from various combinations of *plane* surfaces.

M. *Goupil's* object was to make a series of preliminary experiments with this apparatus, in order to ascertain its stability, the effect of the wind upon such a system, and the resistance to be expected, as well as the sustaining power. He accordingly applied neither motor nor screw, but exposed it to the natural wind when blowing from 18 to 20 ft. per second, say, about 13 miles per hour, at which velocity the resulting air pressure is generally assumed to be 0.85 lbs. per square foot. These experiments took place

in December, 1883, at which season the winds were quite
variable, and the apparatus was anchored by various ropes
so as to prevent it from rising more than 2 ft. from the
ground.

Exposed head on to a wind of 18 to 20 ft. per second,
the body being inclined at an angle of 1 in 10 and the
wings at 1 in 6 (about 10°), this apparatus lifted up clear
of the ground the weight of two men besides its own, making a total of 440 lbs. ! The thrust or end resistance did
not exceed 17.6 lbs. M. *Goupil* tested this several times,
and expresses himself as surprised at the low resistance to
penetration against the wind evidenced by this apparatus,
which was mounted upon two small wheels.

When the wind increased to more than 20 ft. per second
he could no longer control the machine. There being no
stays or guys to the wings, such as are shown in fig. 62,
the apparatus was twisted out of shape, and the wind took
greater effect upon the deformed side. Then a wind gust
occurred ; the efforts of five men were required to control
the apparatus, and one of the wings (constructed with
white pine) was broken.

The inclement season and other considerations of a personal nature prevented M. *Goupil* from pursuing these experiments further at that time. He had gathered valuable
preliminary data, and had caught a glimpse of a very important fact concerning the effect of concavo-convex surfaces, but his own affairs had a more immediate claim
upon his personal attention.

He therefore desisted for a while and allowed the subject to remain in abeyance until he could take it up again,
but he published, in 1884, a very remarkable book, "La
Locomotion Aërienne," in which he advanced a number
of important and new theoretical considerations concerning the solution of the problem of aerial navigation, gave
data concerning the steam-engine, the condenser, and the
various sizes of bird-like aeroplanes which he had designed, and generally evinced such a grasp and comprehension of the question that it seems a marvel that the
book is not more frequently referred to by the French
writers on aviation.

This experiment of M. *Goupil* opens up quite a new field
of inquiry concerning the effects of concave, bird-like surfaces, when exposed to an air current. Calculated by the
data which have been gathered by experiments upon plane
surfaces, the "drift" and total resistance does not seem to
vary greatly from what might be expected, but there is an
enormous, an almost incredible increase of the lifting
power.

Thus there was said to be a total end thrust of 17.6 lbs.

in the apparatus when exposed to a wind of about 13 miles per hour, at which the air pressure would be presumably some 0.85 lbs. per square foot. The angle of incidence of the wings was practically 10°, and we may, without serious error, assume the resistance of the body to have been one-tenth of that due to its mid-section, while that of the edges of the wings (presumably 0.20 ft. in average thickness) would be about one-third of their plane cross-section. As the sustaining surface was 290 sq. ft., we then have, using the table of "lift" and "drift" heretofore given, the following estimate:

RESISTANCE OF THE GOUPIL AEROPLANE.

Drift 10°... ... 290 $\times$ 0.85 $\times$ 0.0585 = 14.42 lbs.
Body.......... 26.9 $\times$ 0.85 $\div$ 10 = 2.28 "
Edge of wings.. 19.7 $\times$ 0.2 $\times$ 0.85 $\div$ 3 = 1.11 "

Total . 17.81 "

which agrees closely with the amount said to have been ascertained by experiment ; but when we come to calculate the lifting force we have :

Lift 10° — 290 $\times$ 0.85 $\times$ 0 332 = 82 lbs.,

while the apparatus is said to have actually lifted 440 lbs., or more than five times as much !

Of course various allowances must be made in considering the results of an experiment carried on in a variable wind, and where so little motion of the apparatus (2 ft.) could be allowed. The thrust may have been measured while the breeze was steady, and the uplift to have occurred during a wind gust, deflected possibly by surrounding objects so as to produce a greater angle than 10° with the wings ; still, in any case, the result of this experiment and also of other experiments by M. *Phillips*, which are to be described hereafter, leads to the inference that much greater supporting power is to be obtained from concavoconvex surfaces than from the flat planes which hitherto have been chiefly proposed for aeroplanes.

This increase in supporting power might indeed have been expected from the theoretical consideration : that the concave lower surface would produce a higher co-efficient of pressure, while the convex upper surface would deflect the current of air impinging at an acute angle thereon, and thus produce a partial rarefaction ; and also from the much stronger practical consideration that *this is the way the wings of birds are shaped ;* and yet very few experiments and proposals seem to have been made with bird-like aeroplanes.

This neglect may possibly be due to the fact that the proportions, the shape, the concavity and the convexity of natural wings differ from each other among the various species, so that the moment that we discard the flat plane, a multitude of combinations present themselves, which may require long and careful experimenting before the best shape for an artificial machine is ascertained.

It is understood, however, that M. *Goupil*, has planned a whole systematic series of such experiments to elucidate this important matter, and that he hopes soon to be in position to carry them on.

In March, 1884, the *Aéronaute* published a paper by M. *De Sanderval*, giving an account of some very interesting experiments, which he had tried with a pair of artificial wings no less than 39 ft. across and 13 ft. wide in the middle. These wings formed an aeroplane, or *rigid* plane of canvas, stretched upon wooden arms, which latter, however, possessed a certain flexibility.

In a first set of experiments, this aeroplane, loaded with ballast to the amount of 176 lbs., was allowed to glide in calm air along a cable 1,300 ft. long, which both supported and guided it, and which was inclined at a slight angle. It was also allowed to drop in still air from a height of 131 ft., and then still further experiments were tried with men riding on the machine when the wind was blowing.

For this purpose the aeroplane and its operator were suspended by a long rope from the middle of a cable, stretched in some cases between two hills and over a ravine, and in other cases between two high masts erected near the sea-shore.

M. *De Sanderval* states that he was attached some 5 ft. above the aeroplane and a little in front of its center of figure, so that by pulling upon four oblique cords he was enabled to shift his weight either forward or back, and to the right or left at pleasure.

When the wind blew and the apparatus was restrained by a head-rope, the effect was much the same as when gliding free in calm air, with, however, the unfavorable difference that when near the ground it was less steady by reason of whirling currents.

In a light wind the apparatus would rise until the suspending rope became horizontal, thus relieving it of its weight-carrying function, and the aeroplane would then oscillate at the pleasure of the operator.

When the wind increased to 18 miles per hour the apparatus would sustain the operator and two assistants.

Subsequently, M. *De Sanderval* gave an account of his experiments to the French Academy of Sciences, and this was reprinted in the *Aéronaute* for November, 1886, with the somewhat uncalled-for comment that " it is a pity that the author should not have stated the time, the place, nor the witnesses, as such extraordinary facts need verifying."

The following are the facts as stated :

My first apparatus consisted in two wings, each 19.68 ft. long, thus giving an aggregate spread of 39.36 ft., by a maximum width of 13 ft. These wings were of canvas, stretched upon bamboos and upon wooden arms. The canvas was divided into a series of parallel sheets or flaps, each 4¾ in. wide, and perpendicular to the dorsal line. They were suitably fastened, and a net was stretched above them, so that they might flap and open upon the upstroke, like the feathers of birds, which oscillate upon the quill which divides them into two unequal portions.

Standing upright upon a light board, and connected by straps to a central spine, I was enabled by thrusts of the legs to develop their maximum effort ; but with this apparatus, which worked quite well, I was enabled to settle but one fact, and that is, that man cannot develop sufficient energy to sustain himself in calm air. I therefore gave up the thought of beating wings.

I then rebuilt the apparatus, transforming the wings into a rigid plane, and replacing the flapping strips by an unbroken canvas.

This apparatus, weighing 99 lbs., and loaded with 176 lbs. of ballast, was caused to glide under a cable 1,300 ft. long, stretched between two bluffs. There was no deflection in the cable when the aeroplane glided across at speed, but the deflection was about 26 ft. when the apparatus was stopped in the middle.

If then released (by tripping a hook) it would at first drop almost vertically : then after the first second it would glide forward at increasing speed, while the rate of vertical fall diminished ; but upon the slightest disturbance in the equilibrium, consequent upon any divergence between the center of gravity and the center of pressure, the inert ballast would aggravate the oscillation, and the apparatus would plunge down to smash. It seemed evident to me that if intelligence were applied to regulate the position of the center of gravity, steady progression would result.

I then suspended the apparatus by a long rope attached in the middle of the cable, and substituted my own person for the ballast. I found that with an intelligent live control the apparatus would oscillate in the wind according to my pleasure. as I have already indicated in a previous communication. The supporting surface of 301 sq. ft. sufficed to sustain a man at a comparatively slow rate of fall, and by a wind of 22 miles per hour it lifted me up with two assistants, and sustained us in the air during the entire period that we kept the holding-back line taut, by maintaining a proper angle of incidence.

The last and more interesting experiment which I attempted was based upon these previous results, and also upon the fact that soaring birds can rise into the air on a helical path, or else maintain themselves a long while at the same altitude without beating their wings, provided always that they possess sufficient horizontal speed as regards the air. I therefore experimented with an apparatus somewhat similar to the preceding, but round in shape, suspended by a vertical rope 650 ft. long,* and caused it to swing around in a circle, so that the suspending rope described in its path the outer periphery of a cone. In this experiment I could feel a notable reaction against my weight, but it required a much longer suspending rope to allow so large an apparatus to swing in a circle of sufficient diameter to permit its gaining the necessary speed, and to manœuvre freely. I believe, however, from the feeling produced upon my mind by the experiment, that I had really taken possession of space within the limits of my somewhat irregular speed, and also, from my observations of soaring birds advancing against the wind on rigid wings, that man can succeed in reproducing sailing flight.

If one had an unlimited height to fall in, affording plenty of time to think and to act, he would probably succeed in guiding himself at will. In calm air man does not possess sufficient energy to sustain himself, but either in a sufficient wind, or with a proper horizontal speed of his own, he finds himself under different circumstances, and derives from the air quite enough supporting power. It is through the operation of this dynamic equilibrium that he will eventually succeed in compassing practical flight.

I caused to be constructed, from manuscript notes furnished by M. *Biot,* a very ingenious apparatus intended to comply with the above conditions, and I experimented with it. This apparatus consisted in two great wings supported on a light carriage, which gained its initial speed by rolling down a long incline covered with an asphalt floor. It rose into the air pretty well, but always with the disadvantage that the experiment could not be sufficiently prolonged to furnish decisive results ; each time upon coming down the apparatus was injured.

It appears to me that a long, vertical rope, such as that previously described, swinging around so as to describe a cone of extended base, must afford greater chances for careful experiment and for eventual success.

The writer has been unable to find any further records of experiments by M. *De Sanderval.* He seems to have been baffled by the lack of means to maintain equilibrium, but even had he possessed the appliances and the skill to bring the center of gravity to coincide with the center of pressure, as often and as fast as the angle of incidence changed, it may be questioned whether he could have acquired, without a very long apprenticeship, that instinctive use of them which constitutes the science of the birds.

---

* Stated at 200 meters ; may be a misprint.

It is inferred from the description that M. *De Sanderval* experimented with plane surfaces, although it is possible that under the action of the wind they may have assumed those concavo-convex shapes which we have seen to obtain with the birds and to be more effective than flat planes. In any case, he is to be commended for having made an earnest if unsuccessful effort *to learn how to soar in a wind like a bird*, the possibility of which performance for man will be further discussed hereafter.

In 1848 M. *Armour*, the author of several papers which will be found in the reports of the Aeronautical Society of Great Britain, patented a flying machine, in which he proposed the use of aeroplanes or wings, oscillating upon springs transversely to the line of motion, these wings being set behind each other as well as superposed. It is not known whether any experiments were tried with this curious device, which seems to be a combination of fixed wings (or aeroplanes) with oscillating wings, but it seems doubtful that it can prove efficient.

There was a second aeronautical exhibition in 1885, under the patronage of the Aeronautical Society of Great Britain, but the total number of exhibits was only 16 as against 78 in 1868.

Among these exhibits the model which attracted most attention was that of M. C. *Ring*, of Denmark, which consisted of an aeroplane with a pair of arched wings, somewhat similar in the front-edge view to the arched wings of the gull and of the albatross. In plan, however, these wings were rectangular instead of the approximately triangular shape which obtains with the birds. These aeroplanes were to act as sustaining surfaces, the angle at which they met the wind being determined by the position of a large flat tail, and the propulsion being furnished by four wing-propellers oscillating beneath the aeroplane, and driven in the model by twisted rubber.

The apparatus was supported by a string fastened vertically above its center of gravity to the crosspiece of a light framework. It propelled itself slowly, but was incapable of free flight, probably in consequence of defective equilibrium.

M. *Ring* also exhibited a model of a gun-cotton engine in which small charges were to be exploded between two pistons, moving in opposite directions in a long cylinder ; but the model was not a working one, and no attempt was made to construct a full-sized engine.

Reference has already been made to a " trunk steam-engine," shown by M. *S. Hollands* at this exhibition. He gave a description of this and of two other types of light steam-engines with which he had experimented, at subsequent meetings of the Aeronautical Society of Great Britain.

The first was a "direct-acting" engine, rotating at high speed *twin* vertical screw fans (right and left) in opposite directions, and a model of this machine, developing $\frac{1}{18}$ horse power, was said to have weighed 6 oz. for the engine and boiler, or at the rate of only 6 lbs. per horse power. It was first intended to generate the steam by burning liquid fuel, but M. *Hollands* subsequently concluded that hydrogen gas, carried highly compressed in a suitable reservoir, and burned with an admixture of twice its volume of air, would prove preferable for lightness and heating efficiency. He estimated that the weight of this type of motor, including not only the engine and boiler, but also the water therein, the fuel-gas reservoir and the driver's stand, would be 11.5 lbs. per indicated horse power.

The other engine was "geared" so as to rotate two right and left fans on *concentric* vertical shafts, one inside of the other, through the intervention of toothed mitre gear. The function of these two vertically superposed fans was to lift only ; a smaller horizontal fan being carried on a prolongation of the crank-shaft, and its thrust aided by the reaction of the exhaust steam ejected through a suitable nozzle. The weight of this engine per horse power is not stated.

Both these arrangements, it will be observed, involved discharging the exhaust steam into the air, and thus wasting some 20 to 22 lbs. of water per horse power per hour, M. *Hollands* not seeing his way to adding an aerial condenser (to recover the steam) in any form, within any admissible limits of weight. He stated that the power necessary was one indicated horse power for every 30 lbs. of the whole weight, so that without a condenser the flight of such an apparatus as he proposed would have been limited by the very small quantity of water which it could lift.

M. *Hollands*, however, made some experiments on the best form of lifting screw-blades, and stated that he had found it advantageous to make the fan blade concave on the driving or lifting side, and that the angle of maximum efficiency was 15° with the plane of motion at the tip, and 30° at the root. The form which he found most efficient was two-bladed ; with the blades narrowest at the tips, slightly concave on the lifting side, the tip slightly drooping, each blade being approximately the shape of an elongated shallow spoon or scoop, and with a pitch equal to about two-thirds of the fan's diameter, giving a mean angle of blade of 22° 30' with the plane of motion. These blades were of thin sheet steel, and their forms will be noted as confirming what has already been stated as to the advantages of the bird-like form of wing. M. *Hollands* said further :

I find another advantage accrues also from the use of these very thin, sharp edged hollow blades—viz., that there is *no appreciable resistance to rotation that does not contribute to lifting effect.* A marked contrast to this desirable quality is presented in the results given by *flexible bladed* fans, constructed to *vary their pitch* automatically, being normally of *coarse* pitch (when still), but decreasing their pitch when rotated, and further decreasing it with increase of speed. Some experiments I made with fans of this description showed an unmistakable loss of power, as compared with the other type above described, due

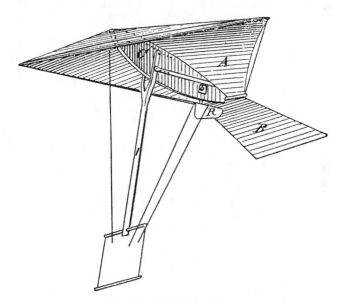

FIG. 63.—BEESON—1888.

apparently to the energy absorbed in deflecting the elastic blades ; which deflection, with a given speed, causes a constant strain and resistance, with no compensating useful effect.

In 1888 *W. Beeson* patented, in the United States, the singular soaring device shown in fig. 63. He had already patented, in 1881, a soaring apparatus consisting of two or more sets of adjustable superposed sails stretched on inverted A frames, which he expected to raise into the air like a kite, and then sail upon the wind, but he apparently abandoned this device in favor of the simpler form shown in fig. 63.

This consisted in a mainsail A and a tail or back-sail B, both of which were supported on a plate or board C, ranging fore and aft. This plate was convexed at its upper edge so that the sail A might extend over, forward and downward to a cross-bar forming the front edge, and thus enclose a head pocket to catch the wind. A forked pendulum-bar, I, was pivoted to the plate C, and it supported at its lower end a trapeze arrangement to carry the operator, who by means of three light cords extending to his hand might alter the angle of incidence of the mainsail A, of the tail B, or of the rudder R. The mainsail and tail being, moreover, connected by an adjustable bar, which caused the mainsail to act upon the tail automatically, so as to maintain the equilibrium at all angles of incidence through the compound lever thus formed.

M. *Beeson* states in his patent that " this machine is self-supporting in a light wind, say, of 10 miles or more per hour, and that when once raised by a kite or otherwise, and cut loose, it will of itself perform the evolutions of a soaring bird and rise to any altitude."

The writer confesses that he has tried the experiment with a small model and has failed ; and so, in the hope that some of his readers may be more fortunate, he has given this account of what seems to be a remarkably simple device—if it will work.

At the Paris Exposition of 1889, Commandant *Renard*, of the French Aeronautical Department, exhibited, in connection with the dirigible war balloon " La France," an apparatus which he had designed some years before (1873) as embodying his conception of a flying machine, and which he termed a " dirigible parachute."

This is shown in fig. 64, and consists in an oviform body, to which is pivoted a couple of standards carrying a series of narrow and long superposed flat blades, intended to sustain the machine when gliding downward through the air.

The dotted lines in the side view indicate the maximum angle of inclination which it was proposed to give to this similitude of a Venetian blind, and it is evident that by setting it at the proper angle, and dropping the apparatus from a balloon, it can be made to travel back against the wind a considerable distance, and also that it may be steered laterally by the addition of a rudder. Beneath the body a sort of skate will be noticed, probably intended to glide over the ground in alighting, or in obtaining initial velocity to rise should a motor be applied ; but the French War Department is reticent concerning its ex-

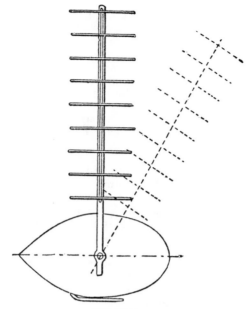

SIDE ELEVATION.

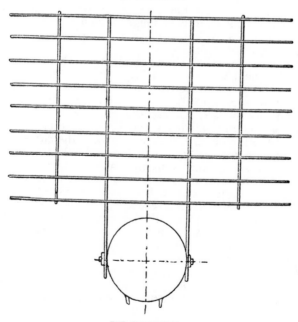

END ELEVATION.

FIG. 64.--RENARD--1889.

periments in aerial navigation, and the writer has been unable to gather any information concerning the working of this apparatus.

It will be noted that Commandant *Renard* proposed to equip this machine with flat blades, thus conforming to the predilection in favor of plane surfaces exhibited by most of the experimenters with aeroplanes already noticed except Captain *Le Bris* and M. *Goupil*, who took a different view as to the best shapes to employ. In point of fact, as already intimated, those who have succeeded in the air, the true experts in gliding, the soaring birds, do not perform their evolutions with plane surfaces. Their wings are more or less convex on top and concave beneath, and are warped surfaces of complicated outlines. It is true that in many cases they do not differ greatly from planes, and the mind of man so strongly tends to the simplification of complicated shapes, that most inventors have assumed that the effect on the air will be practically the same.

Flight is possible with flat planes, as witness the butterfly, the dragon fly, and insects generally, but such creatures are endowed with greater relative power, as already explained ; and, moreover, the elasticity of their wings produces change of shape under action. In the case of the birds, although the outer ends of the feathers are elastic, yet the wing is stiffer as a whole, and the curved surfaces may prove more efficient than planes in obtaining support from the air.

This view seems to have prevailed with Mr. *H. F. Phillips*, for he patented, in 1884, a whole series of curved shapes, intended to be used in conjunction with suitable propelling apparatus for raising and supporting an aerial machine in the air. These shapes were to be utilized in a set of narrow blades arranged at suitable distances apart ; the idea being to defect upward the current of air coming into contact with their forward edges when under motion, so as to cause a partial vacuum over a portion of the upper surface of the blade, and thus to increase the supporting effect of the air pressure below the blade.

These shapes were the result of a series of experiments tried by Mr. *Phillips* in artificial currents of air, produced by induction from a steam jet in a wooden trunk or conduit, and described in London *Engineering* in its issue of August 14, 1885.

A cross-section of the shapes patented will be found on fig. 65, Nos. 1–8. The following table gives the results observed, the last column having been added by myself :

PHILLIPS'S EXPERIMENTS ON SHAPES.

Description of Form.	Speed of Air Current. Feet per Second.	Dimensions of Forms—Inches.	Lift, Ounces.	Thrust, Ounces.	Foot Pounds per Pound.
Plane...	39	16 × 5	9	2.	8.67
Shape 1.	60	16 × 1.25	9	0.87	5.80
" 2.	48	16 × 3	9	0.87	4.64
" 3........	44	16 × 3	9	0.87	4.25
" 4........	44	16 × 5	9	0.87	4.25
" 5.	39	16 × 5	9	0.87	3.77
" 6........	27	16 × 5	9	2.25	6.75
Rook's Wing....	39	0.5 sq. ft.	8	1.00	4.87

The intent of these experiments seems to have been to ascertain the speed of current required to sustain various forms and areas of surfaces, carrying the same weight in a soaring attitude. For this purpose they were exposed to the varying current with their long edges transversely thereto, and they were loaded with a weight applied one-third of the width back from the forward edge, which point was thought to be the center of pressure. These shapes were swung by two wires attached to their front edges, and when they assumed a soaring attitude in the velocity of current required to sustain the weight, the "thrust" or drift was then measured.

The most efficient shape is, of course, that which requires the least expenditure of power, or the smallest number of foot-pounds per pound of weight to keep it afloat, and this is seen to be shape No. 5, which soared with 3.77 foot-pounds per pound, or at the rate of 146 lbs. sustained per horse power, while the flat plane absorbed more than twice as much power.

The comparison would have been more satisfactory if the soaring angles of incidence had been stated. This is given for the plane only as having been 15° by measurement. This agrees fairly well with calculation ; for if the "thrust" is to the "lift" as the tangent of the angle of incidence, then we have $\frac{2}{9} = 0.222 = tang.$ 12° 32'. But all the results obtained were probably somewhat vitiated by assuming that the center of pressure was uniformly one-third of the distance back from the front edge, and therefore applying the load at that point.

We have already seen that this center of pressure varies with the angle of incidence in accordance with Joëssel's law, and the load should have been attached accordingly.

If, for instance, the possible soaring angle were 4°, we should have for the position of the center of pressure, back from the front edge, a distance of $0.2 + 0.3 \sin. 4° = 0.22$ per cent. So that it seems probable that if its load had been applied at 22 per cent. instead of 33 per cent. back from the front edge, the flat plane would have soared at a flatter angle than 15°, and would have shown less "thrust," because the effect of placing the weight so far

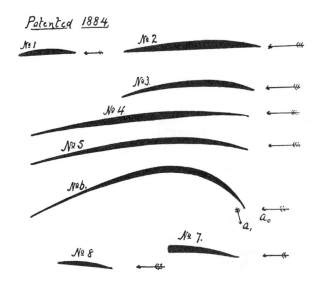

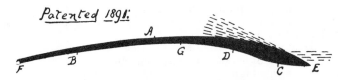

FIG. 65.—PHILLIPS—1884–1891.

back was to tilt the plane unduly, and thus to increase both the angle of incidence and the thrust.

It is not known whether Joëssel's formula applies to curved surfaces ; but be this as it may, it is reasonable to believe that it would be but little modified, so that perhaps the error in locating the center of pressure operated to the disadvantage of the curved forms nearly as much as to that of the plane. We may, therefore, accept the general statement that greater weights per horse power can be sustained in the air with concavo-convex surfaces than with flat planes ; but it seems very desirable that further experiments should be made, for it is quite possible that, in consequence of the loading of the blades at a point differing from the center of pressure, the shapes patented by Mr. *Phillips* are not absolutely the most efficient forms.

It will be interesting, in this connection, to note how these various shapes behaved. It was found that in order to get the maximum efficiency from any given surface, the greatest depth of hollow should be one-third of the total width from the forward leading edge, and that the amount of concavity of the lower surface and the convexity of the upper surface should bear a relation to the speed of the air current. Thus in shapes 1 and 2 the under surface was nearly flat, and the upper curvature not great, while speeds of current of 60 ft. and 48 ft. per second were required respectively to produce a soaring attitude. In shape 3 the curvature was more marked, and the required speed fell to 44 ft. per second. Shapes 4 and 5 were made broader, with a moderate degree of curvature both above and below, and the speeds of current to produce soaring were 44 ft. and 39 ft. per second respectively. Shape 6 was an extreme case, in which the distinguishing features of the experiments were purposely carried to excess ; for when impinged upon by a current of air of 27 ft. per second in the direction of the arrow a_0, it was seen (by a fine attached ribbon) that there was an induced current flowing outward in the direction a_1.

Shapes 7 and 8 were used to demonstrate that the impinging air is deflected upward by the forward part of the upper surface, and that a partial vacuum results in the after part ; they were not loaded with weights, and when exposed to a current of air of sufficient velocity, coming in the direction of the arrow, they rose into the position shown in the figure.

In 1890 Mr. *Phillips* patented an aerial vehicle in which these curved surfaces were applied to an apparatus similar to the " dirigible parachute" of Commandant *Renard*, except that there were to be two (or more) series of curved blades behind each other at suitable distances apart. They were to be attached to an elongated body, which he indicated might be of fish shape, and, say, 30 ft. long. The cross-blades, which he termed " sustainers," might be 15 ft. long, 6 in. wide, and 2 in. apart, so many being superposed as to furnish the required supporting air surface. Each set of " sustainers" was to be held in place by a number of vertical bars of angular form, so as to offer the least resistance to the air.

The propelling power was not indicated specifically, save the general statement that it should be "suitable," but a rudder was located at the top of the front series of curved blades, being affixed to a spindle bar terminating below (at the body) with a lever arm. A shifting weight was also provided, capable of being moved across the body, transversely to its line of motion, in order, when moved to either side, not only to depress it, but, by the resistance of the air acting on the surface of that weight, to check forward motion on that side, and thus cause the machine to describe the curve required.

The patent drawings show the vertical standards carrying the blades as being rigidly attached to the body instead of being pivoted thereto, as in the case of Commandant *Renard's* device, and hence the angle of incidence of the machine could not be conveniently varied in order to rise or to descend ; but it is probable that Mr. *Phillips* has long since remedied this defect, for he is understood to have been continuously experimenting, although the results attained have not as yet been published.

He apparently concluded that he had not developed the best shape in 1884, for he patented, in 1891, the form shown at the bottom of fig. 65. In this, the upper side *A* of the blade was made convex, as formerly, but the after portion of the lower side of the blade was made concave, as shown at *B*, while the curvature of the forward portion of this lower side was in the form of a reverse curve consisting of a convex curve, *C*, at the forward edge, followed by a concave curve, *D*. He states in his patent :

"The particles of air struck by the convex upper surface *A* at the point *E* are deflected upward, as indicated by the dotted lines, thereby causing a partial vacuum over the greater portion of the upper surface. The particles of air under the point *E* follow the lower convex and concave surface *C D* until they arrive at about the point *G*, where they are brought to rest. From this point *G* the particles of air are gradually put into motion in a downward direction, the motion being an accelerating one until the after edge *F* of the blade is passed. In this way a greater pressure than the atmospheric pressure is produced on the under surface of the blade."

Mr. *Phillips* indicates that such blades may be of wood, 12 ft. in length and 6 in. in width, from the leading edge *E* to the rearward edge *F*, but further experiment led him to make these blades still narrower, and he finally constructed an experimental machine which was tested in the early part of 1893, and has been described in various English journals, notably in *Engineering* of March 10 and May 5, 1893, the latter issue containing four illustrations, which were reproduced in the AMERICAN ENGINEER for June, 1893. From these various publications the following description of the Phillips experimental machine is compiled.

Instead of providing two series of curved blades, one behind the other, there was but one set, approximately as shown in fig. 64. The apparatus looks like a huge Venetian blind with the slats open. There are 50 of these slats or "sustainers" 1½ in. wide and 22 ft. long, fitted 2 in. apart in a frame 22 ft. broad and 9½ ft. high. The sustainers have a combined area of 136 sq. ft. ; they are convex on the upper surface and concave below, the hollow being about $\frac{1}{16}$ in. deep. The frame holding the sustainers is set up on a light canoe-shaped carriage, composed principally of two bent planks like the two top streaks of a whale boat, and being 25 ft. long and 18 in. wide, mounted on three wheels 1 ft. in diameter, one in front and two at the rear. This vehicle carries a small boiler with compound engine, which works a two-bladed aerial screw propeller revolving about 400 times per minute. The fuel used is Welsh coal. There is said to have been no attempt to provide exceptionally light machinery. The weights of the various parts of the machine are, approximately : carriage and wheels, 60 lbs. ; machinery with water in boiler and fire on grate, 200 lbs. ; sustainers, 70 lbs. ; total weight, 330 lbs. The machine was run on a circular path of wood with a circumference of 628 ft. (200 ft. diameter), and to keep it in position (preventing erratic flight) wires were carried from various parts of the machine to a central pole, as in the *Tatin* experiments heretofore described. Still further to control the flight, which there is no means of guiding, as the machine is not of sufficient size to carry a man, the forward wheel is so balanced that it never leaves the track, and therefore serves as a guide, carrying some 17 lbs. of the weight, the remainder being on the hind wheels.

On the first run 72 lbs. dead weight were added, making the total lift 402 lbs. As soon as speed was got up, and when the machine faced the wind, the hind wheels rose some 2 or 3 ft. clear of the track, thus showing that the weight was carried by the air upon the Venetian blind sustainers. A second trial was made with the dead weight reduced to 16 lbs. and the circuit was made at a speed of about 28 miles per hour (2,464 ft. per minute), and with the wheels clear of the ground for about three-fourths of the distance. That the machine can not only sustain itself, but an added weight, was demonstrated beyond all doubt, even under the disadvantages of proceeding in a circle, with the wind blowing pretty stiffly.

It is stated in the journal *Iron* that the boiler is a cylindrical phosphor-bronze vessel 12 in. in diameter and 16 in. long. The fire grate area is 70 sq. in., and the fuel Welsh coal. The engine is compound, having cylinders 1¾ in. × 3⅝ in. × 6 in. stroke, fitted with ordinary slide valves. The working pressure of steam is 180 lbs. per square inch. The propeller is 6 ft. in diameter and 8 ft. pitch, with a projected blade surface of 4 sq. ft. The machine was also moored by a stern rope in which a dynamometer was inserted, and on the engine being run at full speed the dead pull was 75 lbs.

If the latter figures be correct, then the power developed was $75 \times 2,464 \div 33,000 = 5.6$ horse power, and the weight carried per horse power was $402 \div 5.6$, or 72 lbs. per horse power, which is inferior to the 110 lbs. per horse power carried by M. *Tatin's* apparatus, and probably due to the increased resistance produced by the frame which holds the sustainers.

Mr. *Phillips's* experimental machine neglects any provisions for maintaining equilibrium in full flight, or for rising and alighting safely. Those he may add later ; but whether he does or not, he is entitled to great credit for having been among the first experimenters who have tested concavo-convex surfaces instead of adhering to plane surfaces, and who have thus drawn attention to what may prove to be a very important line of inquiry.

Almost all scientific experiments in air have hitherto been tried with planes, and such few formulæ as have been proposed are based upon the effect on flat surfaces. It is probable that such formulæ—those of Smeaton, Duchemin, Joëssel and others—will be found to need modification, either in form or in constants, when applied to curved surfaces. In such case the tables of "lift" and "drift" heretofore given herein will either need recalculation for each specific curved shape, or require the application of a variable coefficient, as exemplified in the calculations of the power expended by the pigeon as heretofore given. In any case it seems very desirable that further scientific experiments be made on concavo-convex surfaces of varying shapes, for it is not impossible that the difference between success and failure of a proposed flying machine will depend upon the sustaining effect (with a given motor) between a plane surface and one properly curved to get a maximum of "lift."

Fig. 66 represents a kite-like aeroplane proposed by M. *de Graffigny*, a French aeronaut, and the author of several works upon aerial navigation. This apparatus was to consist of a kite 46 ft. across, with its fabric surface capable of bagging to a certain extent, and attached

to a longitudinal frame, as shown, which was to be trussed
both above and below. In front, a stiff triangular head
was to be affixed, and an adjustable horizontal tail was to
be placed in the rear. Between these, a boat-shaped body
containing the machinery and aviators was to be swung

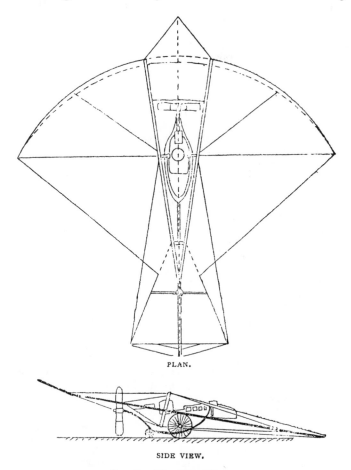

PLAN.

SIDE VIEW.

FIG. 66,—GRAFFIGNY—1890.

on trunnions and attached to the frame. In front of this
car a two-armed screw was to rotate, and behind the car
a vertical steering rudder was to be placed, above the surface of the kite.

M. *de Graffigny* estimated that the power required to
drive the apparatus was in the proportion of one horse

power for every 110 lbs., and he proposed the use of liquefied carbonic acid gas, which he states to weigh but 55 lbs. per horse power, including the motor, the recipient and a supply for several hours. This, of course, was a mere makeshift, a reservoir of power for experiment, and not a prime mover ; inasmuch as the whole apparatus was to weigh but 396 lbs. and to have sufficient sustaining surface (some 1,300 sq. ft.) to come down like a parachute, should the motor break down while in the air. The screw was to be 6 ft. in diameter and 10 ft. pitch, and its shaft was to remain constantly horizontal (this being the object of hanging the car on trunnions), so that the position of the propeller should be independent of the angle of incidence of the sustaining surface, in accordance with the theory of the designer.

M. *de Graffigny* states that he experimented with a model of this apparatus in 1890. The screw was rotated some 300 turns per minute by a skein of twisted rubber threads weighing, in the aggregate, 1.1 lbs., and producing 1,085 foot-pounds in 2½ minutes, or at the rate of 7.23 foot-pounds per second, which proved quite insufficient to give to the apparatus (mounted on three wheels, the foremost of which was adjustable) the velocity necessary to cause it to rise upon the air. The designer expresses himself as unable to state what would be the result with a full-sized apparatus.

It will be noted that this proposal resembles a number of others which have already been described. It is probable enough that the best form for sustaining a given weight and for propelling it with a minimum of surface and of power, or for maintaining equilibrium, were not selected ; but M. *de Graffigny*, in the book* in which this design is incidentally described, strongly advocates the kite principle generally, as the one most likely to lead to success in devising a flying machine, and in learning how to manage it in the air.

This will have occurred to many readers, and it may be interesting to them to inquire as to what has been published upon past experiments with kites, a subject upon which the writer has found distressingly little on record.

Among the first, if not the very first, to call attention to the fact that the study of the kite as a means of obtaining unlimited lifting and tractive power had been unduly neglected was Mr. *Wenham*, who, in his celebrated paper on "Aerial Locomotion," published in 1866, described briefly some very interesting experiments with kites, and who has kindly furnished the writer with some additional

---

* "Traité d'Aérostation." H. de Graffigny, 1891, p. 189.

particulars. Mr. *Wenham* states that his principal summary of facts was taken from a little book, styled the "History of the Charvolant, or Kite Carriage," by Mr. *George Pocock*, of Bristol, England, who also published a small work on "Aeroplastics," both of them, unfortunately, now having become very rare.

The experiments described took place more than half a century ago, and the purpose of the inventor was not to evolve a flying machine, but to provide a floating observatory to serve in warfare, or to drag wheeled vehicles over land.

The apparatus was, in fact, a huge kite, of suitable size to carry the intended weight, with a chair swung just below, and so rigged that by tightening or slackening the different cords which held it, the wind would meet it at any angle desired, and the apparatus would rise or fall, or could be made to swing a considerable distance to one side or the other. It was so arranged that in case the cords broke, it would act like a parachute, and thus insure safety.

The following quotation, descriptive of the experiments, was given by Mr. *Wenham* in his paper :

" While on this subject we must not omit to observe that the first person who soared aloft in the air by this invention was a lady, whose courage would not be denied this test of its strength. An arm-chair was brought on the ground ; then, lowering the cordage of the kite by slackening the lower brace, the chair was firmly lashed to the main line, and the lady took her seat. The main brace being hauled taut, the huge buoyant sail rose aloft with its fair burden, continuing to ascend to the height of 100 yards. On descending she expressed herself much pleased with the easy motion of the kite and the delightful prospect she had enjoyed. Soon after this another experiment of a similar nature took place, when the inventor's son successfully carried out a design not less safe than bold—that of scaling, by this powerful aerial machine, the brow of a cliff 200 ft. in perpendicular height. Here, after safely landing, he again took his seat in a chair expressly prepared for the purpose, and, detaching the swivel line, which kept it at its elevation, glided gently down the cordage to the hand of the director. The buoyant sail employed on this occasion was 30 ft. in height, with a proportional spread of canvas. The rise of the machine was most majestic, and nothing could surpass the steadiness with which it was manœuvred ; the certainty with which it answered the action of the braces, and the ease with which its power was lessened or increased. . . . Subsequently to this an experiment of a very bold and novel character was made upon an extensive down, where a wagon with a considerable load was drawn along, while this huge machine, at the same time, carried an observer aloft in the air, realizing almost the romance of flying.

"It may be remarked (continues Mr. *Wenham*) that the brace lines here referred to were conveyed down the main line and managed below ; but it is evident that the same lines could be managed with equal facility by the person seated in the car above ; and if the main line were attached to a water-drag instead of a wheeled car, the adventurer could cross rivers, lakes, or bays with considerable latitude for steering and selecting the point of landing, by hauling on the port or starboard bracelines as required. And from the uniformity of the resistance offered by the water-drag, this experiment could not be attended with any greater amount of risk than a land flight by the same means."

The reader may perhaps inquire whether there was not some risk that the kite should run away with the wagon when the wind freshened ; but Mr. *Wenham* further explains that the kite attached to the "charvolant" or chariot was provided with a smaller "pilot," or upper kite, which was sufficient to support the "draft," or lower kite, when it was relaxed or allowed to float edgewise, on the wind. The "draft" kite had two cords, one attached well forward, and the other attached well aft, running through rings to keep the cords together. If the aft cord was slacked off by the driver of the chariot, the "draft" kite floated edgewise on the wind, and the wagon stopped ; but by pulling on the aft cord the kite could be made to face the wind absolutely, and to produce the maximum of draft.

Mr. *Wenham* also mentions in his paper Captain *Dansey's* kite, for communicating with a lee shore, as described in Vol. XLI. of the "Transactions of the Society of Arts." This was made of a sheet of holland fabric exactly 9 ft. square, and, as stretched by two spars placed diagonally, spread a surface of 55 sq. ft., the remarkable fact about its performance being that in the experiment about to be quoted this surface of 55 sq. ft. sustained no less than 92¼ lbs. The quotation is as follows :

"The kite, in a strong breeze, extended 1,100 yards of line ⅝ in. in circumference, and would have extended more had it been at hand. It also extended 360 yards of line 1¾ in. in circumference, weighing 60 lbs. The holland weighed 3½ lbs., the spars, one of which was armed at the head with iron spikes for the purpose of mooring it, weighed 6¾ lbs., and the tail was five times its length, composed of 8 lbs. of rope and 14 lbs. of elm plank, weighing together 22 lbs."

This latter kite seems to have been provided with a tail to steady it in the air, and in considering the bearing of such experiments upon possible flying machines, it is preferable to select those upon tailless kites, sailed with one single line, for it is easy to maintain the stability if several restraining cords be used. Mr. *Wenham* has kindly furnished to the writer the particulars concerning a tailless kite, or, rather, series of superposed kites, patented in Great Britain in 1859, by *E. J. Cordner*, an Irish Catholic priest, who designed the apparatus to save life in shipwrecks, and who preferred to arrange hexagonal disks of fabric (stretched upon three sticks), above each other on the same line, so that they would all pull together. The operation was to be as follows :

When a sailing-vessel had struck, which almost in every case occurs by the ship being blown on a lee shore, a common kite was to be elevated in the usual way from on board the vessel. When enough cord had been paid out to keep the kite well suspended, the end of the cord on board was to be attached in a peculiar manner to the back of another and larger kite (without tail), and the second kite was then to be suffered to ascend. The end of the suspending rope was to be attached in a similar manner to the back of another and still larger kite, and the process to be repeated until enough elevating and tractive power was obtained, when a light boat or basket with one occupant was to be fastened to the kite line, the latter being paid out until the occupant reached the shore and alighted, when by means of a light running line, extending from the ship to the person ashore, it was deemed easy to haul the basket back and forth as many times as necessary to rescue the passengers and crew.

It is not known whether this ingenious method of saving life without extraneous aid was ever used in a case of actual shipwreck, but it was tested by transporting a number of persons purposely assembled on a rock off the Irish coast, one at a time, through the air to the main land, quite above the waves, and it was claimed that the invention of thus superposing kites so as to obtain great tractive power was applicable to various other purposes, such as towing vessels, etc.

Many proposals have been made at various times and in various countries to utilize kites in life saving, but none seem to have come into practical use. Such attempts may have suggested to Mr. *Simmons* (the English aeronaut) the experiments which he is said to have tried, in 1876, of gliding downward under such buoyant sails.

The only accounts which the writer has found of these experiments are given in the *Aéronaute* for April, and for November, 1876. The apparatus of Mr. *Simmons* is described as consisting of a huge "pilot" kite 49 ft. high and 49 ft. wide, with another kite below, still larger. The pilot kite was first to be raised, and to carry up the second ; the two were to be adjusted to the breeze, and the aeronaut was to be suspended in a car, and allowed to ascend 200 or 300 yards. Then by adjusting his weight by means of guy lines, so as to obtain a proper angle of incidence, the apparatus was said to glide downward to the ground, being slightly dirigible through the guy lines, and to be arrested by the bystanders seizing a dragging guide rope.

Mr. *Simmons* is said to have been fairly successful with his experiments in England, but to have failed to repeat the feat at Brussels, Belgium. In the latter case it was claimed that there was not sufficient wind, but steadiness of breeze would be more important. The surfaces operated with seem to have been very large—some two to three square feet per pound in order to alight gently ; but such extent of surface is so unmanageable in a gusty wind as probably to have led to the abandonment of the experiments.

The exploit is feasible, and would prove useful in experimenting with various shapes and extent of surfaces, but such experiments should be tried with areas more nearly corresponding to the proportions which exist in soaring birds, and the operator should invariably alight in water until he has learned how to manage his apparatus.

In July, 1880, M. *Biot* exhibited to the French Society for Aerial Navigation an ingenious kite, invented by himself, which sailed without a tail and possessed great stability under all conditions of wind.

At the top of a flat plane of elongated elliptical shape two hollow cones were affixed, one on each side. The base or large end of these cones faced the wind, and the other end or point was slightly truncated, so as to leave an opening through which the wind could blow, and, by the action of the streams or columns of compressed air thus created, counteract any tendency of the plane to tip to one side or to the other. This provided for the lateral stability on the same principle as in the well-known Japanese kite, in which the side-pockets catch the wind and maintain the equipoise.

The fore and aft equilibrium was provided for by affixing a rotating screw at the lower end of the plane, pivoted on its central line. This screw had two vanes of coarse pitch, and was free to rotate under the impulse received from the wind. It spun around with great speed when the kite was raised, and obviated any need of the usual tail by performing the same steadying office. It prevented any oscillations, without impeding the rising of the kite, and maintained it perfectly steady in all winds.

It was not agreed between the French aviators whether this effect was due to the action of the vanes, making an angle with the sustaining plane, as in the case of *Pénaud's* "planophore," or to "gyroscopic" action, but when the screw was omitted the kite swayed about, while when the screw was rotating, its twirling and tremor could be felt through half a mile of string, and the kite remained perfectly upright and steady.

M. *Biot* carried on quite a series of experiments with this apparatus. In the kite which he used the elliptical plane was 15 in. high, the two cones at the side were each 8 in. in diameter at the base by a height of 8 in., while the screw was 12 in. in diameter, its two vanes being each 1¾ in. broad.

The experiments were carried on in winds varying from 13 to 33 miles per hour, and the kite was found to be steady under all conditions, the only difference being in the height to which it would rise. When the wind blew from 13 to 18 miles per hour, 4,900 ft. of cord were paid out, the kite remaining at this distance during two hours. On other occasions, with stronger wind, as much as 6,500 and 8,200 lineal feet of cord were paid out, and the kite mounted so high that it passed through several strata or currents of wind of varying direction, as was conclusively proved by the fact that the restraining cord assumed a sinuous attitude when the full height was gained, and instead of approximating to a straight line or a regular curve, as usual, the line became serpentine in form, thus indicating that different trends existed in the various strata of air.

In one instance the kite, with 2,600 ft. of cord paid out, advanced against the wind and mounted directly over the head of the operator. This was attributed to an ascending trend in the wind, for the kite still tended to rise vertically and to advance against the wind, although the plane was horizontal, and the cord, now greatly bowed by the wind, tended to drag the apparatus backward. This attitude continued but a short time, when, the trend of the wind having apparently changed, the kite settled back to its original position, flying at an angle of 40° to 60° with the horizon.

M. *Biot*, who was an old experimenter with kites (having as early as 1868 been lifted up from the ground by a large apparatus of this kind), found the gyroscopic stability of the arrangement which has just been described so satisfactory, that he thereupon designed, in connection with M. *Dandrieux*, a full-sized aeroplane on the same principle, calculated to carry up a man. This design was submitted early in 1881 to a special committee of the French Society for Aerial Navigation, but this committee seems to have hesitated in recommending its construction, and no record has been found by the writer of its having been built or experimented with about that time.

When, however, the publication and discussion of M. *Mouillard's* "L'Empire de l'Air" had directed fresh attention to the soaring of birds on rigid wings, and given grounds for the belief that man could utilize the wind in the same way, M. *Biot* constructed in 1887 a soaring apparatus in the shape of an artificial bird 27 ft. across, and weighing 55 lbs., with which he hoped to reproduce the manœuvres of the sailing birds.

It is known that a number of very interesting experiments were tried with this apparatus, but the writer has been quite unable to find in print, or to obtain from correspondents, a description of the machine or a record of its trials. He merely knows that these trials were many, and that on one occasion M. *Biot* suffered a tumble which was not encouraging to further experiment, but no account of them is to be found in the *Aéronaute*.

It is to be hoped that a full narrative may yet be given to the public of the results of experiments which must have been most instructive for other aviators who contemplate imitating the birds.

In 1882 M. *Jobert* exhibited before the French Society for Aerial Navigation the model of a proposed apparatus designed by himself, in order to test the possibility of imitating the manœuvres of the soaring birds, as described by M. *Mouillard*. This aeroplane was to be hinged and jointed, so that it might be folded up like an umbrella for convenience in transportation, or opened out and stiffened by sliding bars in order to make the wings rigid. With this M. *Jobert* proposed to experiment on various areas of surfaces in proportion to the weight, and to test the efficacy of both fixed and adjustable sustaining surfaces. He does not seem to have met sufficient encouragement to carry out his design, for the writer has been quite unable to learn that he ever completed a full-sized aeroplane capable of sustaining a man.

Having begun where M. *Biot* terminated—*i e.*, with the design for a soaring apparatus—M. *Jobert* next turned his attention to kites, and proposed in 1887 the apparatus shown in fig. 67, which he termed a "rope-bearing kite," designed for establishing communication with wrecked vessels. It consisted of a hollow truncated cone *C*, under which was rigidly connected a kite *P*, from which depended two light lines terminating in a ring, the latter carrying a light cord steadied by the drag *D*. The object of this arrangement was to ensure a rapid and certain connection with the shipwrecked mariners, who, by seizing the light cord, could at once haul down the kite and thus gain access to the main carrying rope, with which to haul aboard the usual life-saving cable. This carrying rope was fastened to a bridle attached to the top cross-stick of the kite, and to the top of the cone at V, which arrangement was claimed to produce perfect stability, and to ensure that the apparatus should travel straight back in the line of the wind without rising to any great elevation. In order to regulate the height, the angle of the plane could be varied by means of a light string (not shown), extending from the lower cross-stick to the carrying rope and fastened by a hook in one of a series of loops.

The sustaining plane was, like the cone, formed of calico, in which hems were turned at the top and at the two sides, in order to form cases for the sticks of the frame,

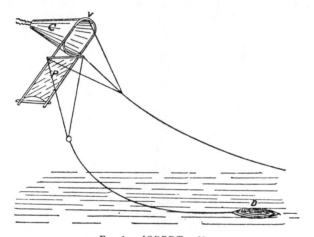

FIG. 67.—JOBERT—1887.

the lower edge of the kite being left uncased, in order to produce bagging and consequent increase of lifting power. At the small end of the cone a couple of thin metallic tongues were fastened, which, thrown into vibration by the wind rushing through the cone, produced a howling sound which might notify the shipwrecked sailors of the approach of the apparatus, and the whole arrangement, as will readily be perceived, was quite cheap and readily rigged up or folded away, no matter how large it might be.

The writer does not know whether this apparatus ever came into practical use. It has here been figured in order to show how a cone can be applied to a kite in order to impart stability to the latter, but the arrange-

ment would need to be greatly modified in order to admit
of its utilization in an aeroplane, so as not unduly to increase the resistance to forward motion.

In 1886 and 1887 M. *Maillot*, a French rope-maker,
tried quite a series of experiments with the kite represented in fig. 68. This was constructed of poles and canvas, in the shape of a regular octagon ; it measured 775
sq. ft. in area, about 32 ft. across, and weighed 165 lbs.
It had neither balancing head nor tail, and was so poised
by the bridle of attachment that the center of pull corre-
sponded to a point only one-third of the distance back
from the front edge, or to a spot, therefore, decidedly forward of the center of pressure, at the comparatively coarse
angles (30° to 60°) usually assumed by kites. This angle
of incidence it was intended to regulate by a cord, attached to the rear edge and carried to the seat swung
beneath the kite for the operator, who might then, by
hauling in or paying out this cord, regulate the angle of
incidence and cause the kite to rise or to fall. This was
intended to furnish the longitudinal stability, while (there
being no provision for automatic lateral equilibrium) the
side oscillations caused by the varying intensity and directions of the wind were restrained by side ropes attached
to the kite and handled by men standing on the ground.

In the first experiments (May, 1886) M. *Maillot* was dissuaded from ascending beneath the kite, and he therefore
substituted for his person a bag of ballast weighing 150
lbs., tied just below the seat. The kite was raised by
first securely anchoring the main rope, which was 800 ft.
long, and then lifting up the front edge so that the wind
might sweep under the surface ; upon which the kite rose
to such height that the bag of ballast swung some 30 ft.
above the ground, where M. *Maillot* and two assistants
managed the two side ropes and the tail cord (not shown
in the figure), which latter regulated the angle of incidence by depressing or raising the rear of the kite.

Allowing 33 lbs. for half the weight of the main rope,
it was estimated that the apparatus sustained, on this
occasion, an aggregate weight of 348 lbs., or in the proportion of 2.23 sq. ft. per pound. The wind was variously estimated at 15 to 22 miles per hour ; but as this
speed was not measured, nor the pull upon the various
ropes ascertained, while the angle of maximum incidence
was merely guessed at, as about 45°, no accurate computation can be made of the various reactions. The kite
was easily controlled by the three men, hauling or paying
out the two side ropes and the tail cord, but it plunged
about with the varying intensity of the wind, and in one
of the oscillations so produced the bag of ballast was

whipped about and broke the rope by which it was suspended.

M. *Maillot* repeatedly experimented with this and other

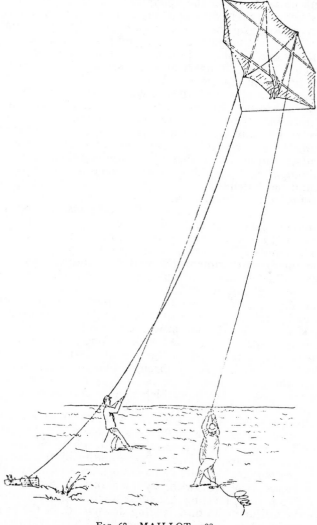

Fig. 68.—MAILLOT—1887.

kites (but smaller) on the same principle during the year 1887. He states that he succeeded in sustaining as much as 594 lbs., but whether he ever went up himself beneath the kite the writer has been unable to ascertain. There would have been little or no risk in doing so, provided the wind was steady and strong, for it is evident that the three lines carried to the ground would give almost complete command over the apparatus, but then such a performance would have taught very little toward the management of an aeroplane free in the air. Changes were made from time to time in the modes and points of attachment of the various ropes, and the endeavor seems to have been directed to the discovery of some arrangement by which automatic equilibrium could be secured, under all conditions and varying velocities of wind, without the use of a tail. From the discontinuance of the experiments it is inferred that they did not succeed, and the writer attributes this failure (if failure it was) to the employment of a *single* rigid *plane ;* for it will be remembered that M. *Pénaud* obtained a stable kite, on the principle of his " planophore," by adding to the upper pair of planes a second set, inclined at a slight angle to the first, the effect of which was to regulate the incidence.

On the same principle, M. *Barnett*, whose proposed aeroplane has already been noticed, obtained stability with a tailless kite many years ago, by shaping the plane like a laundry " flat-iron," cutting out a portion of this from the rear or broad end, and adjusting the band so obtained at an angle with the rest of the surface, so that the kite would fly steadily.

M. *Copie*, on the other hand, obtained partly the same effect by inserting a hemispherical pocket in the body of the kite, but this did not prove quite satisfactory until an opening was cut in the apex, on the same principle as the hole which is provided in the top of a parachute, after which the wind, rushing through the pocket, produced much the same effect as in the *Jobert* regulating cone ; but the device is not one which can be profitably applied to an aeroplane in forward motion.

Upon the whole, M. *Maillot's* kite was rather crude, and decidedly inferior to *Pocock's* " charvolant," heretofore described, in which the pilot kite might be used to regulate the carrying kite. The stability of the *Maillot* arrangement could probably have been improved, and the side ropes dispensed with, by breaking up the surface into two planes, forming a diedral angle with each other, like the attitude of a bird gliding downward, or the same effect might have been partially produced by providing the plane with a keel.

Very good results with central keels have been obtained by M. *Boynton* with his various forms of " Fin " kites, which are now sold in the shops. They consist of a plane, to which is affixed at right angles a "fin" or keel located in the lower part of the kite, and raised slightly above its surface. They fly without a tail, with a steadiness depending somewhat upon the form of the main bearing surface, and seem to afford a good opportunity for further experiment as to the shape of greatest stability ; for keels have been frequently proposed for aeroplanes, in which they will produce less resistance to forward motion than obtains with other arrangements, but few seem to have tested how such keels should be applied.

These remarks chiefly apply to *plane* rigid kites, and to the various adjuncts and forms which have been tried in order to confer stability upon a main plane surface sustaining the weight ; but still better results have been obtained with flexible surfaces, and it seems not improbable that this is the arrangement which will give the greatest amount of stability to a kite, by producing automatic adjustment to the wind's varying intensity.

As an example of such action may be mentioned the " Bi-Polar" kite of M. *Bazin*, who experimented with it in 1888. It consists of a main sustaining surface like a boy's " bow" kite, or practically the same in shape as the kite surface in fig. 66. The frame is composed of two sticks, one of them a flexible rod at the head, bent to a bow, and the other a main central spine at right angles, to which the bow-strings are fastened. The peculiarity of the " Bi-Polar" kite is that this central spine is also made flexible, and that to its lower end (projecting some distance below the supporting surface of the kite) three triangular fins are attached, just like the tail of a dart, omitting one fin. This arrangement obviates any necessity for a tail and confers automatic stability, for the lateral equilibrium is obtained through the elasticity sideways of the main surface or head, which is blown back by the pressure to a convex surface with a diedral angle, which angle varies in accordance to the violence of the wind, while the longitudinal equipoise is likewise maintained by the balancing pressures on the head and on the fins, as the flexible spine yields more or less to the breeze. The kite is thus made stable in both directions, and flies steadily without a balancing tail. M. *Bazin* sailed it with two strings, one attached at the top and the other at the bottom of the main sustaining surface ; these strings were both carried to the ground, and attached at each end of a stick of equal length with the vertical distance which they spanned at the kite, and with this stick in his hand the operator could vary the angle of incidence. This was intended to secure measurements of this angle of incidence in connection with the pull, but the results thus obtained have not as yet, to the writer's knowledge, been published.

Even better results can be attained with the "Malay" kite, which is in shape a lozenge, composed of two flexible sticks crossing each other at right angles. The cross or horizontal stick is the longest, being preferably 1.14 times the length of the upright stick, and fastened to the latter at a point 0.18 of its length below the top; a string is then carried (in notches at the ends of the sticks) around the periphery of the resulting lozenge, and this is covered with paper or with muslin in the usual way. This surface, when impinged upon by the wind and restrained by the bridle, is bent back by the pressure and adjusts itself to the varying irregularities of the breeze, the kite flying without a tail with great steadiness and rising to great elevations.

M. *Eddy*, of Bayonne, N. J., who has been constantly experimenting with kites during the last few years, and who is recognized as an expert in such matters, prefers the "Malay" kite to all others. He has improved it by so fastening the cross-stick and tying its outer ends as to produce a slight initial convexity, which is further increased by the action of the wind, and which materially adds to the steadiness of the flight. With this arrangement M. *Eddy* has succeeded in causing a single kite to ascend to a height of 2,400 ft. with 3000 ft. of line, and then bringing it to the zenith directly over his head, or even a little back of his hand, where its attitude strongly suggested the advance of the soaring bird against the wind. Upon a previous occasion he had succeeded in attaining a height of 4,000 ft., with a string of five kites flying in "tandem"—that is to say, each kite attached by a string of its own to the string of the preceding kite already raised, so as to take up the slack or sagging of the line, and thus enable the upper kite to rise to an altitude otherwise unattainable. This performance seems to suggest an easy way for the exploration of the upper air by the Weather Bureau, for by affixing to the upper kite self-registering instruments (thermometer, barometer, hygrometer, etc.), or, preferably, by connecting such instruments (and an anemometer besides) electrically with recording instruments on the ground (through a series of fine wires insulated in the kite string), observations of the conditions prevailing aloft can be easily obtained. The French have lately been making such observations by means of "free balloons" of medium size, and they are said to be of material assistance in forecasting the weather; the records obtained from the top of the Eiffel Tower showing that even at that moderate height coming changes in wind and in temperature are indicated several hours in advance of their prevalence at the ground.

The same principle of obtaining stability without a tail, by means of an elastic frame, can be applied to other forms than the " Malay" or the " Bi-Polar" kites, but it requires a good deal of delicate adjustment and balancing. It has been done with the common hexagonal form of kite by M. *C. E. Myers*, of Frankfort, N. Y., the aeronautical engineer who furnished and operated the balloons and kites by means of which the recent (1891) rain-compelling experiments were tried in Texas.

It will be remembered that the explosions intended to produce rain were in some cases produced by exploding dynamite suspended below kites, and fired by electricity. In providing for this, M. *Myers*, who has for several years been conducting systematic experiments with kites, evolved some very interesting facts, and he has published part of his experience in the *Scientific American Supplement* (No. 835) for January 2, 1892, from which the following is extracted :

The originating cause of my interest in kite flying is aerial navigation, and by successive steps I have adapted kites to fly without tails, to fly with considerable weight attached, and, finally, to fly without the restraint of the usual kite-string ; and, rising higher and higher, finally to disappear miles in height and miles away on the verge of the distant horizon.

Theoretically, there should be no difficulty in attaining these results. Practically, there is as much difficulty as with a child learning to walk or a youth learning to manage a bicycle. In a word, it is the art of balancing. . . .

Theoretically, the kite should be light or possess much surface with little comparative weight. It should balance at the flattest possible angle, nearly horizontal, and its surface should be widespread, like the wings of a soaring bird. As a fact, I have obtained the best results with this model, but had great difficulty at first to induce it to fly at all, and was finally forced to attach a compromise tail—not a kite tail, but a bird-like tail, which, being flexible, vibrated or undulated with the vertical oscillations of the kite, and thus acted as a propeller, so that this kite actually moved against the wind. . . .

The most practical form of kite for general purposes seems, however, to be the six-sided. Those created by me as part of my apparatus for the Government rainfall expedition in Texas were composed of an X, formed of two spruce sticks, each 6 ft. long, tapering, with a top section of $\frac{1}{4}$ in. $\times \frac{1}{4}$ in. and bottom section of $\frac{1}{4}$ in. $\times \frac{1}{2}$ in. tacked flatwise together with a very small pin-nail, and bound with hemp cord at the joining. Five in. below this crossing (which was about 2 ft. from the top) was a similar piece of timber, but 14 in. shorter, and tapering each way, placed crosswise of the X, horizontally, so as to form a

5-in. triangle, which stiffened the frame more than if all crossed at one point. The outer end of each stick was creased with a knife and notched around, so that a hemp cord passed first through the crease and was "half-hitched" around each stick to prevent splitting. The kites were covered with red calico, pasted on tight, and bits of cloth were also pasted across the sticks where the kite-strings attached. These strings were attached as long loops—one loop to the top sticks about 6 in. from their tips, one loop to the two bottom sticks about 30 in. from the bottom, and one loop to the cross-bar about a foot from each end. All these loops were then gathered together and drawn through one hand as the kite lay on the ground, held in place by one foot on its crossing, and being adjusted carefully and equally to draw from *a point somewhere midway between the cross-stick and top*, best attained by trial, were then tied together.

The kite was thus rather stiff and light at top, *elastic* and heavier at the bottom, and suspended at a point *above* its center of gravity and center of surface. To the loop at the bottom was usually hung a narrow strip of cloth to afford greater steadiness in supporting the kite's burden of dynamite to be exploded. I have been thus particular to describe minutely this construction because many have written me for this information.

The first trial kite flown at Midland, Tex., escaped. I had built it all myself, as a model, and it had drawn up one ball of hemp twine, and an assistant was holding the string preparatory to running out another ball when the cord parted at a flaw, and the kite flew into space. When last seen with a glass, it was estimated to be about 3 miles high and 8 or 10 miles away, a fading red dot in the distance. . . .

In ordinary light winds this kite floats well, is steadier than many other kinds I have tried, and would seem to be well adapted for photography. If hung very near its top, it is prone to advance upward and forward against the wind, till over and beyond the party holding the string, and literally floats on the air as if propelled by its fluttering triangular section at or near the bottom of the kite.

The accidental escape of this kite exhibited a very interesting example of partial "aspiration," and it is understood from additional information, kindly furnished by M. *Myers*, that he succeeded in reproducing this effect on several occasions. The kites were hung, after considerable experiment, so that they floated nearly flat on the air, with as little tail as possible, and sometimes none at all. They rose upon a light breeze, and drew away as long as the string was let out. When checked or pulled, they rose higher and higher until quite overhead, when the string had to be released. If suitably balanced the kite then rose still higher and drifted back, *but not as fast as the wind blew*, its rearward flap vibrating more or less, and *making its action a progressive one relative to the wind*, thus producing "aspiration" with respect to the breeze. A long string, or small weight at the end of a shorter string, was sufficient to keep it balanced, so that it might remain up for hours and go floating out of sight. The possibility of this progressive action against the wind without loss of height (or of "aspiration") has been strenuously denied, and yet it is easily explained if, instead of assuming the wind to blow horizontally, as we generally do, we consider that it has at times a more or less ascending trend, this being a not unusual condition over the sunbroiled plains of Texas. It is clear, from the description of the mode of attachment of the string, that its weight when released would tilt the kite forward, so that the plane would point below instead of above the horizon. In this position the direction in which the "drift" is exerted would be reversed—that is to say, the horizontal component of the pressure, instead of pushing backward, would be pulling forward, and thus *become a propelling force against the wind*, provided, of course, that the latter still exerted its pressure on the *under side* of the kite. Thus an upward trend in the breeze of but 3° or 5°, operating against a kite inclined forward 2° below the horizon, would be sufficient to cause it to advance relatively to the wind, somewhat as a vessel "close hauled" advances against the breeze which furnishes its motive power. In point of fact, therefore, that which has herein been termed the "drift" may act upon a *plane* surface, as a force pushing backward or propelling forward, according as that *plane* is inclined to the front *above or below the horizon ;* but in the latter case there needs be an ascending trend in the wind in order to produce a sustaining pressure on the under side, for otherwise the horizontal wind would strike the upper surface of the *plane* and press it downward instead of upward. The effect may be quite otherwise with concavo-convex surfaces.

Ascending trends of wind are by no means rare, as abundantly proved by published observations since M. *Pénaud* called attention to the many causes which must produce such trends. This was shown in a very able paper on "Sailing Flight," which was published in part in the *Aéronaute* for March and April, 1875, but which, unfortunately, was left unfinished. M. *Pénaud* demonstrated that such winds must necessarily result from even moderate undulations of the ground (and therefore *a fortiori* from mountains or deep valleys), from natural or artificial objects acting as wind breaks, from the meeting of air currents flowing in different directions, or even from the heating effect of the sun. He doubtless expected to show, in the portion of the paper remaining unpublished, that an upward trend of $\frac{1}{9}$ to $\frac{1}{6}$ (from 6° to 10°) in the wind was quite sufficient to enable a sailing bird to progress against the breeze by inclining his aeroplane so. that the horizontal component of the pressure would have a forward direction, while the wind still acted on the under side ; for we have already seen in computing the foot-pounds expended by a 1-lb. pigeon in gliding, that with a speed of 40 miles per hour and an angle of incidence of 3° the "drift" will be 0.05647 lbs., while the body resistance and that of the edges of the wings together will be 0.05555 lbs., and that at 5° (30 miles per hour) the "drift" will be 0.08892 lbs., and the resistance of the body and edge of wings will be 0.03124 lbs., so that in both these cases the "drift" (calculated even with the coefficients which have been obtained with planes, and which are known to be inferior to those to be expected from concavo-convex surfaces) is sufficient, if directed forward, to overcome the resistances and to give to the sailing bird a forward impulse ; this reversal in direction of the "drift," as previously explained, occurring when the plane becomes inclined so as to point forward below the horizon.

Since *Pénaud's* day a great many observations have confirmed the frequent prevalence of both ascending and descending currents. Aeronauts, more particularly, have noted that the atmospheric currents follow the undulations of the ground, causing their balloons to subside upon approaching a valley, or to rise when nearing a cliff or a mountain. They have also inferred, from the fact that they have found butterflies a mile or more above the earth while sailing over table lands, that these trends are frequent in such regions, although their effect upon the balloon is less immediately noticeable than in mountainous countries, where the angle of ascent often is 45° or more. In such broken countries very curious observations have been made as to the invariable prevalence of steeply ascending winds in certain well-defined localities when the wind blows from a particular quarter ; such, for instance, as the observations of M. *Mouillard* in the Lybian chain near Cairo, and those of M. *Bretonnière* in the vicinity of Constantine, Algeria, where certain zones or gaps of ascending winds seem to exist, which the sailing birds utilize to gain elevation by circling. There they congregate in crowds, forsaking the rest of the sky, and spirally mount on rigid wings, until they have gained sufficient altitude to carry them toward any point which they may want to reach in descending.

It is probably in sub-tropical regions that such phenomena are most numerous and permanent ; but the reader, who is accustomed to thinking of the wind as blowing horizontally, may be quickly edified by watching the smoke issuing from a tall chimney even in northerly climates. This smoke will be seen at various hours, or on various days, to trend either upward or downward or with exact horizontality, as may depend upon the undulations of the great atmospheric waves which are produced by the impinging upon each other of the currents flowing and crossing at various altitudes ; or if the observer have the good fortune to be in the regions inhabited by the sailing birds, he may satisfy himself as to the similar atmospheric undulations which are constantly taking place, even in a

FIG. 69.—SIMPLEST CHINESE KITE.

perfectly flat country, such as the plains of Texas or the sea beaches of Florida, by liberating bits of down or threads of smoke from the same spot at various times or days. He will also observe the local ascending currents permanently produced by a mere wind break, such as a belt of trees facing the inflowing sea breeze. He may satisfy himself (by attaching light strips of bunting or bright-colored threads to the tops of those trees) that the breeze is deflected upward just over their upper branches, and he will then understand why these spots constitute the favorite haunts of the sailing birds when the breeze is light. He will see the soarers for hours gliding back and forth and back and forth on pulseless wings, just above the top of the wind break formed by these belts of trees, evidently utilizing the ascending current to patrol the adjoining beach while awaiting, with no labor, whatever food may be brought by the incoming tide, or an opportunity of eating it undisturbed.

It is not intended here to convey the impression that ascending trends of wind are absolutely necessary for sailing flight. The writer has seen the feat performed many times, when every test seemed to prove that the current was absolutely horizontal ; but it then seemed to him that on such occasions the equilibrium was more difficult to maintain, and that the bird had to bestow greater attention upon the nice adjustments required to preserve his balance and to produce " aspiration" when the wind varied in intensity and direction ; just as an acrobat experiences greater fatigue in walking a tight rope, through the attention and care expended to avoid falling, than in walking many times the same distance on the ground, where no particular care is required to preserve the balance. It is probably because of such relief from all cerebral strain that the soaring birds seem to sail with less care and with far greater steadiness when-

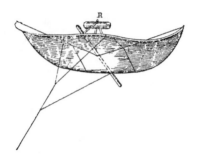

FIG. 70.—CHINESE MUSICAL KITE.

ever they are utilizing an ascending current. They are then easily and safely sustained, and so mechanical does the performance seem that some observers have expressed the opinion that they then sleep on the wing. There is no doubt, moreover, that ascending trends of wind enable the creatures to soar in lighter breezes than would otherwise be possible, and when the faint morning wind first begins to blow, many of the sailing birds will be seen congregated just above wind breaks, while the other parts of the sky are vacant.

But to return from this digression, occasioned by the feat of " aspiration" performed by M. *Myers's* kites, it will be discerned that the principle of flexibility alluded to confers stability upon the well-known *Japanese kite,* specimens of which are now to be found in almost all toy shops. This kite flies without a tail, the frame being so light and elastic that the surface adjusts itself constantly to the irregularities of the breeze. The side pockets catch the wind, and by springing back of the medial line form a diedral angle which confers lateral balance, while the flexibility up and down confers longitudinal equilibrium. The same principle is exemplified in the upward bending of the extremities of the feathers of birds in flight, which doubtless adds much to their stability, and, indeed, so universally is this principle illustrated by all creatures which navigate fluids, that Dr. *Amàns*, in a work upon the locomotive organs of fishes,* lays it down as an axiom derived from physiological considerations, that an aeroplane of rigid form is *contre nature*, or in direct antagonism with all the inferences to be drawn from the observation of creation.

The Japanese are expert kite-flyers, and have produced many shapes besides that which has been above alluded to. They are said to use kites as weather vanes, and to have hitching posts in their gardens to which the device is almost permanently affixed. Indeed, it is said that these kites sometimes remain 8 or 10 consecutive days up in the air—an astonishing achievement to European and American kite fanciers, who seldom succeed in keeping their apparatus up more than a few hours. The explanation is probably to be found in the greater regularity and permanence of the air currents in the regions of trade winds, and these too are the regions where the soaring birds are most numerously found, probably because they are there sure of a sustaining breeze every day, through the use of which they may evade the fatigue of flapping flight.

The various forms of the *Chinese kites* are even more numerous than those of the Japanese, and most of the tailless kind are said to depend upon the same principle of flexibility for their equilibrium. It would not at all be surprising to find, should a stable aeroplane be hereafter produced, that it has its prototype in a Chinese kite ; but the writer has discovered very little information in print upon the subject ; the following article, translated from *La Nature* by the *Scientific American* and published in its issue of March 24th, 1888, being perhaps the best available :

One of our correspendents in China, Mr. Huchet, at present in Paris, has had the kindness to have made for our purposes, by a skillful Chinese manufacturer, a series of models representing the different types of kites used everywhere in China, Annam, and Tonkin, and which the same gentleman has been obliging enough to bring to us in person.

Fig. 69 represents the simplest form of these kites. Its frame is formed solely of a stiff bamboo stick, *A B*, and two

---

* Comparaison des organes de la locomotion aquatique, P. C. Amans.

slightly curved side rods, $C D$ and $E F$. To this frame is pasted a sheet of paper, which is somewhat loose at the extremities $C E$ and $D F$, where, under the action of the wind, pockets are formed that keep the affair bellied and in an excellent position of equilibrium. Our engraving shows the mode of attaching the strings that serve to hold it. Kites of this kind are usually about 3 ft. in width.

Fig. 70 shows the appearance of the musical kite, so called because it is provided with a bamboo resonator, R containing three apertures, one in the center, and one at each extremity. When the kite is flying, the air, in rushing into the resonator, produces a somewhat intense and plaintive sound, which can be heard at a great distance. This kite is somewhat like the preceding, but the transverse rods of its frame are connected at the extremities and give the kite the aspect of two birds' wings affixed to a central axis. This kite sometimes reaches large dimensions—say 10 ft. in width. There are often three

Fig. 71.—CHINESE BIRD KITE.

or four resonators placed one above another over the kite, and in this case a very pronounced grave sound is produced. Mr. Huchet informs us that the musical kite is very common in China and Tonkin. Hundreds of them are sometimes seen hovering in the air in the vicinity of Hanoi. This kite is the object of certain superstitious beliefs, and is thought to charm evil spirits away. To this effect it is often, during the prevalence of winds, tied to the roofs of houses, where, during the whole night, it emits plaintive murmurs after the manner of Æolian harps.

Among ingenious fancies of the Chinese is their bird kite, fig. 71, the frame of which is made elastic. The thin paper attached to the wings moves under the action of the wind and simulates the flapping of the wings. This kite is sometimes 3 ft. in length.

The most curious style of Chinese kite is the dragon kite, fig. 72. It consists of a series of small elliptic, very light disks formed of a bamboo frame covered with India paper. These disks are connected by two cords which keep them equidistant. A transverse bamboo rod is fixed in the long axis of the ellipse, and extends a little beyond each disk. To each extremity of this is fixed a sprig of grass which forms a balancing plume on each side. The surface of the foremost disk is slightly convex, and a fantastic face is drawn upon it, having two eyes made of small mirrors. The disks gradually decrease in size from head to tail, and are inclined about 45° in the wind. As a whole, they assume an undulatory form, and give the kite the appearance of a crawling serpent. The rear disk is provided with two little streamers that form the tail of the kite. It requires great skill to raise this device.

This last device resembles in arrangement the multiple disk kites for life saving of the Rev. Mr. *Cordner*, already described, and suggests that the superposition of kites affords a good field for experiment. There is a limit in

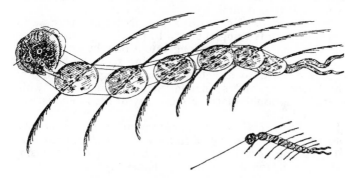

FIG. 72.—CHINESE DRAGON KITE.

size beyond which the increasing leverage will so add to the required strength and weight of the frame as to make a kite unduly heavy as well as unwieldy,* and superposition naturally suggests itself for experiments intended to test the efficacy and equilibrium of kite aeroplanes. There will be many practical details to work out in devising the best mode of attachment of such aeroplanes with each other, so that all surfaces may pull together and yet counteract the effects of wind gusts, so that experiments with kites seem to offer the readiest, quickest, and least expensive method of working out this part of the problem.

The attention of experimenters is specially called to the form of kite shown in fig. 70. It resembles in shape

---

* The largest kite on record is said to belong to a Japanese gentleman, and is 50 ft. × 45 ft., weighing 1,700 lbs. Its frame is composed of 350 pieces of wood.

and attitude those of the soaring birds, which, as already remarked, perform their manœuvres with peculiarly curved and warped surfaces, and it will be seen hereafter that the nearest success in compassing gliding flight hitherto obtained—that of M. *Lilienthal*—has been achieved with just such surfaces.

Inventors seem to have bestowed but little attention upon kites, less than a score of such devices having thus far been patented in the United States. These patents chiefly cover various methods of making the frames to fold, so that the kite may be more portable, while but few inventors seem to have considered how the stability may be increased. Among these latter may be mentioned Mr. *Clarke* (No. 96,550), who proposes the insertion of a spring on one of the three cords which compose the bridle. By the yielding of this spring the angle of incidence of the kite may vary somewhat with the varying velocities of the wind, and thus diminish the perturbations.

· Mr. *Maddans* (No. 121,056) proposes a kite with a convex surface, this being obtained by providing a stick across the top, which stick is sprung into a bow by attaching its ends to each other ; but this bowing seems to have been chiefly devised to attach a flapping tongue, rotating on the bowstring, and so making a drumming noise, while there is no doubt that the convexity of the kite must add to its stability.

Mr. *Thompson* (No. 225,306) patents a reversible convex or concave kite, with a frame like that of an umbrella ; but nothing is said of the equilibrium or of dispensing with a tail, the object being, apparently, to provide for convenience in carrying.

Mr. *Colby* (No. 354,098) provides for the stability by inserting in the middle surface of a kite a wind bag rearwardly projecting, which is distended by the breeze and prevents the kite from darting. This is virtually the same device as that of Mr. *Copie*, already mentioned, which was found to require a central opening to allow the escape of the air when experimented in large dimensions. It is evident that such a device, if applied to a navigable aeroplane, would largely increase the resistance to forward motion ; but this might be minimized by making such wind pockets very shallow, and inserting a large number in the aeroplane. The experiment may be worth trying by kite fanciers.

While several forms of folding frames for kites have been patented by inventors, few seem to have been designed to act as parachutes also. This has been accomplished recently by Mr. *Moy* in a very simple way (British patent No. 1,916, A.D. 1892) by providing the folding frame with a central hub, to which a trapeze bar may be suspended when such a kite is used for conveying passengers or for exploration. By using two lines, the angle of incidence may be controlled, and the kite be made either to raise a weight or to descend slowly to the ground as a parachute.

As already intimated, the writer has found singularly little on record concerning kites, and that little bears but slightly upon the important question of the stability of aeroplanes. It may be for lack of more thorough search that only fragmentary information has been gathered. Kites are supposed to have been invented 400 years before the Christian era by Archytas, a resident of Smyrna (where the flying of kites remains a national sport to this day), and the Asiatics have always been and are now the great kite experts of the world. It is, therefore, not improbable that search in books of travel or inquiries addressed to Orientals might elicit information bearing directly upon the flying machine problem ; and it is much to be desired that some competent person shall undertake to write a critical account of kite experiments as well as of the kites of all nations, and of the influence of form as to stability and sustaining power. There is a large collection of Chinese kites in the National Museum at Washington, and it would certainly be interesting to have an account of the various principles exemplified and of the behavior of the various shapes in the air.

At the annual meeting of the American Association for the Advancement of Science, held in Buffalo, N. Y., in 1886, a paper on The Soaring Birds was read by Mr. *Lancaster*, then of Chicago, which paper attracted great interest and attention.

Mr. *Lancaster*, in the hope of surprising the secret of the birds, had the pluck, in 1876, to exile himself to the wilderness of Southwestern Florida, on the Gulf coast, near the Everglades, and there to remain for five consecutive years watching the sailing of the master soarers. He published some of his observations and deductions in the London *Engineer*, in the reports of the Aeronautical Society of Great Britain, and in the *American Naturalist ;* but the subject was by no means exhausted, and his description of the phenomena observed was so interesting to the members of the association, few of whom had ever seen a soaring bird at close range, that they demanded to hear more upon a subsequent day.

Unfortunately for Mr. *Lancaster*, upon the latter occasion he attempted to give a mechanical and mathematical explanation of the performances which he had previously so well described, and his theory was so plainly erroneous that he was subjected to harsh ridicule and criticism. He had witnessed some remarkable feats of " Aspiration," he had attempted to reproduce them artificially, but he was clearly wrong in his expounding of the mystery, and his critics did not properly discriminate between the statements of observed facts and the attempted explanation.

The principal issue, however, was made concerning some attempts to imitate soaring action, which Mr. *Lancaster* claimed to have successfully made in Florida, and which he unwisely declined to exhibit at Buffalo. He had described them in his paper as follows :

I constructed floating planes which, for lack of a better name, I have termed " effigies," and which are an example in point. I have made scores of them. They would draw into the breeze from the hand and simulate the soaring birds perfectly, moving on horizontal lines or on an inclination to a vertical. They would float in the best winds with neither ledge, rough front surface nor rear curve, if very nicely adjusted ; but one of this construction I never induced to pass beyond the limits of vision, as the equilibrium was so very delicate that a little inequality in the wind current would capsize it.

There is every probability that such an experiment would invariably fail if tried in any but a perfectly steady sea breeze, an inflowing curent of air with peculiar conditions ; but it does not follow that the action described is impossible, for if we presume that current to have an upward trend (and the writer knows, of his own knowledge, that such upward trends are not rare in sea breezes), we can readily see that an aeroplane, tipped forward so as to point below the horizon, may be both sustained and " aspirated," or possess a forward component of pressure, so that it may advance against the wind, and rise at the same time, like the kite of M. *Myers*.

The equilibrium is, of course, excessively delicate, and hence the requirement for a sea breeze ; for a local homogeneous mass of air flowing from the water over the land to replace the rising quantity heated by the sun ; but it is erroneous to suppose that the experiment described by Mr. *Lancaster* involves a mechanical impossibility. It is, doubtless, difficult to repeat it successfully, because it requires a combination of peculiar conditions ; but the soaring birds are daily performing the feat, and apparently in horizontal winds.

In order that those who are favorably circumstanced may test the matter experimentally, the following description of the device is copied from a paper of Mr. *Lancaster*, published in 1882.

Take a stick of wood 1 in. square and 18 in. long, and point one end. Slit the other end 3 or 4 in., and insert a piece of stiff cardboard 6 in. wide and 1 ft. long. This will represent the body and tail of the bird. Fasten on both sides near the pointed end a tapering stick 2 ft. long, with the outer ends slightly elevated, and fasten to these and the body a piece of cardboard 10 in. wide and 2 ft. long. Have the tail vertical instead of horizontal, as in the bird. Round off the outer rear corners of the wings for 3 or 4 in. The imitation of the natural bird is now complete. There is no need of exactness, as the air you are to try it in will be an unknown quantity, and it may just suit the shape you make. An indispensable part is now to be added, which is to preserve the equilibrium and is not used by the natural bird. A tapering stick, say 1½ in. wide, ⅜ in. thick at the top, ⅜ in. square at the other end, and 18 in. long is used. This piece is to be securely fastened by a small bolt through the upper end of the body piece, about 5 in. from the front end. It must be capable of adjustment by allowing the lower end to swing front and back through say 40°. To the lower end is fastened a muslin bag which will hold 2 lbs. of shot. Expose the effigy to a breeze of from 3 to 20 miles an hour, from as high a situation as it is possible for you to obtain, by holding it by the pendant stick near the body. Adjust the weighted stick forward or back, and add or subtract shot until the effigy has a tendency to spring from your hand against the current of air, when it may be released at a moment of greatest steadiness of breeze.

I have made hundreds of these toys, with all kinds of success, but have never yet succeeded in getting one to travel beyond the limits of vision. They have proceeded directly against the breeze for 500 yards, and obliquely, up or down, or to right or left, within those limits, when they would lose their balance and come down. Sometimes almost any kind of one that was presented to the air would float creditably, while at others none would succeed. The pendant weight for maintaining equilibrium, though the best I have ever devised, is far too sluggish for perfect work. The momentum of this weight prevents the best results, for, if a succession of puffs of wind upon the same wing should occur quickly together, the weight would swing far enough, in obedience to the impulse given, to capsize the effigy. Such a succession of puffs is sure to occur, sooner or later, at each trial. These toys operate long enough, however, to prove the purely mechanical character of flight, and serve to materially strengthen the theory.

The writer may confess that he has tried this experiment several times under special instructions furnished by Mr. *Lancaster*, but that he has never succeeded in floating one of these "effigies" so that they would advance against the wind. Others have, to his knowledge, tested the matter, and had no better success, yet it is not rational to say that the feat is impossible, for it is very clear that if the wind have an ascending trend, and the "effigy" be slightly tipped toward the front, the horizontal component of air pressure will drag it forward, while the vertical component will sustain or elevate it, as already explained. It is probably because of the uncertain prevalence of ascending trends that Mr. *Lancaster* complains that sometimes almost all these toys would succeed in simulating soaring, and sometimes none at all.

In 1888 Mr. *Lancaster* moved to Colorado, where he has been experimenting with a view to the solution of the problem of soaring flight ; and in the *American Naturalist* for September, 1891, he gave an account of some of his experiments, with the conclusions which he deduced therefrom. The following extract contains an account of a remarkable occurrence. He says :

I can produce true soaring flight in natural wind, with a plane exceeding 2 lbs. to a square foot of surface, whenever I wish to do so and can obtain wind strong enough for the purpose. During the past three years I have made about 50 planes [aeroplanes ?] of various shapes and sizes, and from 25 lbs. to 400 lbs. in weight. These planes are not set free in wind, but used in the experimental cases above described, but with rigid rods in place of the parallel wires. These rods run in large rings and have a cross-head at their outer ends allowing the plane to run to the front until its edge rests against the rings. In the best trial the parallel [with the plane ?] component is neutralized at 10° from horizontal, far exceeding my expectations derived from observations of the birds, their angle of obliquity being rarely over 5°.

On a few occasions these planes accidentally escaped me in time of highest wind, and were ruined at once for all purposes excepting firewood, in each case being a loss of two or three months' work, and playing havoc with my finances. One that I valued particularly plunged to the front in a violent blast of wind with force sufficient to tear out the rings. It rose into the air, gradually higher and higher, until an elevation of at least 3,000 ft. was attained, when some part of the device giving away, it lost equilibr'um and plunged through the air, striking the earth about 2½ miles from the starting-point, and 1,000 ft. higher than that locality. Another mile would have carried it to the summit of the Flat Top Mountains. It was in the air about three hours, and I walked beneath it during its flight. Its course was directly against the highest wind I have experienced during my residence here. At times it did not progress, but went higher. It weighed 110 lbs., and had been well balanced for experimenting on surface manipulation. There was no lesson taught in this flight, the birds having been doing the same thing for a long time. It was an interesting spectacle to look at ; so is a large bird in the same act. I presume Mr. Darwin's provisional solution would apply to this plane as well as to the condors ; but I am trying to explain the actual mechanical activity of both.

The best effects produced were with a plane of 400 lbs.

weight and 80 sq. ft. of surface. In a wind that would be rightly termed a gale, arising about midnight, this plane was thrown about 7° from horizontal. It ran to the front against the rings at 10°, where the entire parallel component was neutralized, and at 7° it hugged the rings with a force that required a backward pull of 15 lbs. to detach it.

This plane would make a splendid navigator, and I would have no hesitation in trusting myself to it, when steering, equilibrium and alighting or stopping items had been worked out. I mean to say that it would navigate wind. I am now just entering on a course of experiments in calm air.

This very interesting case of "aspiration" may have been produced by the same cause as in the case of M. *Myers's* kite—*i.e.*, an ascending trend of wind ; but certainty concerning this depends upon the shape of the surface. Mr. *Lancaster* writes of it as a "plane ;" but as he mentions also the "front ledge" and the "rear curve," the surface operated upon by the wind was probably a more or less compound surface, for which there is no specific name, but which may be described as an aeroplane. If it was shaped like those of the soaring birds, then "aspiration" might occur with a horizontal wind, but the equilibrium would be very unstable, and, as Mr. *Lancaster* points out, the steering, alighting, and stopping would be the important points to work out.

Among the most systematic and carefully conducted series of experiments that have ever been made in the direction of artificial flight are those of Herr *Otto Lilienthal*, of Berlin, Germany, a mechanical engineer and constructor, and a prominent member of the German Society for the Advancement of Aerial Navigation.

The general position that he maintains, and in pursuance of which he has made his more recent experiments, is that bird flight should be made the basis of artificial flight. Dexterity alone, as he maintains, invests with superiority the native denizens of the air, and, therefore, man, if he possessed sufficient skill, might participate in flight. He evidently believes, like M. *Mouillard*, that for the soaring birds ascension is the result of the skillful use of the power of the wind, and that no other force is required ; and, therefore, that to imitate them no engines or other external sources of power are needed, but that all the necessary apparatus consists of properly constructed sustaining surfaces skillfully operated.

Herr *Lilienthal*, instead of first flying at conclusions, began by a systematic analysis of the problem, verified by experiments, which latter were carried on by himself and his brother, *G. Lilienthal*, during a period covering nearly 25 years, and he published in 1889 a book on

" Bird-Flight as the Basis of the Flying Art," * in which he gave the result of his investigations.

From a review of this remarkable book in the *Aéronaute* for January, 1892, the following account of its contents has been prepared.

Herr *Lilienthal* seems to have begun by observing the sailing of various sea birds following vessels at sea, and of the stork, an expert soarer, which inhabits Germany ; he drew the conclusion that *plane* surfaces present undue resistance, and that success in artificial flight is only to be expected from concavo-convex sustaining surfaces ; a belief which, as we have already seen, was also entertained by *Le Bris, Beeson, Goupil, Phillips*, and others.

He declares that the laws of air resistances and reactions which, unfortunately, are as yet but imperfectly known, form the whole basis for the "technique" or actual performance of flight, and that the shapes and methods of birds so completely utilize these laws and offer such appropriate mechanical movements that failure must follow if they be discarded.

Herr *Lilienthal's* experiments were in great part directed toward an investigation of the resistances and reactions of air, and the power necessary for flight. One of these consisted in suspending himself from a spar projecting from a house and operating a set of six wings opening and closing like concave Venetian blinds, through which he measured the lifting effects of wing strokes performed with the muscles of the legs, so that the step of each foot would produce a double stroke of the wing. The weight of the operator and wings combined was 176 lbs., and they were counterweighted with 88 lbs. suspended to a rope passing over two pulleys. With some practice he was enabled, by operating the wings with the pedals, to lift himself 30 ft. from the earth, thus proving that he obtained, through his mechanism, wing power sufficient to lift the remaining 88 lbs.—a very excellent performance, and much in excess of most of those hitherto described in this review of Progress in Flying Machines.

This and other experiments, together with a consideration of the power to be obtained from the wind, convinced him that artificial flight was accessible to man, aided by considerably weaker motors than have generally been thought indispensable, and, indeed, under favorable circumstances of wind, with no motor at all.

Herr *Lilienthal*, therefore, carefully analyzed the shapes and methods of the living birds and the exact proportions

---

* " Der Vogelflug als Grundlage der Fliegekunst," Von Otto Lilienthal, Berlin, 1889.

of their concavo-convex surfaces. He went into this in detail, and finally formulated in his book the following conclusions :

1. The construction of machines for practical operation in nowise depends upon the discovery of light and powerful motors.

2. Hovering or stationary flight without forward motion cannot be compassed by man's unaided strength. This mode of flight would require him to develop, under the most favorable circumstances, at least 1.5 horse power.

3. With an ordinary wind man's strength is sufficient to work efficiently an appropriate flying apparatus.

4. With a wind of more than 22 miles per hour, man can perform soaring or sailing flight by means of adequate and appropriate sustaining surfaces.

5. A flying apparatus, in order to operate with the greatest possible economy, must be based, both in shape and proportion, upon the wings of the large, high-flying birds.

6. The sustaining wing surface may be from 0.49 to 0.61 sq. ft. per pound of weight.

7. Sufficiently strong apparatus can be built of willow frame and stretched fabric, so as to provide a sustaining surface of 107 sq. ft., with a weight of about 33 lbs.

8. A man provided with such an apparatus would have an aggregate weight of 198 lbs., and would then have 0.55 sq. ft. of sustaining surface per pound, or about the proportions of large birds.

9. Experiment must determine whether the most advantageous shape be that of birds of prey and of waders, with broad wings and spread out primary feathers, or that of sea birds, with narrow wings tapering to a point.

10. If the broad wing be adopted, the wings of an apparatus with 107 sq. ft. of sustaining surface would needs be of 26.25 ft. spread, with a maximum width of 5.25 ft.

11. If the narrow wing be adopted, a surface of 107 sq. ft. would need a spread of 36 ft. with a maximum width of 4.60 ft.

12. The application of an additional bearing surface, as a tail, is of minor importance.

13. The wings must be curved in transverse section so as to be concave on the under side.

14. The depth of flexure should be one-twelfth of the width, in order to correspond with that of birds' wings.

15. Experiment must determine whether greater or lesser flexure will prove preferable for larger wing surfaces.

16. The framing and spars of the wings should be at the front edge so far as possible.

17. A sharp cutting edge should terminate this framed front edge if possible.

18. The flexure should be parabolic, the greater curvature being to the front and flatter to the rear.

19. The best shape of flexure for large surfaces must be determined by experiment ; also what preference is to be given

to those shapes which produce the least resistance to forward motion at flat angles of incidence.

20. Construction must be such as to admit of the rotation of the wing upon its longitudinal axis, which rotation will best be obtained, in whole or in part, by the pressure of impinging air.

21. In flapping flight the inner wide portion of the wing should oscillate as little as possible, and serve exclusively in sustaining weight.

22. The propulsion to maintain speed should be obtained by up-and-down beats of the wing tips or of the primary feathers, the forward edge being depressed.

23. In flapping flight the widest portion of the wing must also co-operate in the up stroke in order to sustain weight.

24. The wing tips should encounter as little resistance as possible on the up stroke.

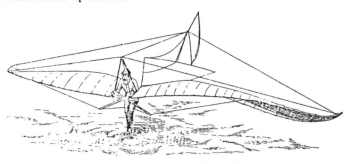

FIG. 73.--LILIENTHAL—1891.

25. The down stroke should be in duration at least six-tenths of the time occupied by the double stroke.

26. The wing-tips alone need oscillate ; that portion of the wing which merely sustains may remain rigid, as in soaring flight.

27. If only the wing-tips oscillate they should not be articulated, as this would dislocate them ; moreover, the transition to the up stroke should be as gentle as possible.

28. In order to beat a pair of wings, man must employ his extensor muscles, and this not simultaneously, but alternating each side, so that each stroke of the foot shall produce a double stroke of the wings.

29. The up stroke may be produced by the pressure of the air under the wings.

30. The energy of the air pressure under the wings may be partly stored in a spring so as to restore the power on the down stroke, and thus produce economy in work done.

Such are the principal considerations which must be observed in the application of the theories herein expounded. . . .

Governed by these considerations, equipped with much preliminary experiment and analysis, Herr *Lilienthal* put his theories and conclusions to practical test, in the summer of 1891, by undertaking a series of experiments with a pair of curved wings designed for soaring alone—that is, to serve as sustaining surfaces and not for flapping or propulsion.

The following account of these experiments has been furnished by Mr. *George E. Curtis*, of Washington, D. C., who has also obtained from Dr. *C. Kassner*, of the Meteorological Institute at Berlin, the very graphic photographs from which the engravings have been made.

The *Lilienthal* apparatus is shown in fig. 73, and consists of a pair of extended bird-like wings, incurvated from front to back on parabolic lines, and sinuous in the direction of their lengths. The area of sustaining surface, as at first constructed, was 107 sq. ft., but it was diminished in the course of numerous changes and remodellings to 86 sq. ft. There was, as will be observed, a horizontal tail and a vertical rudder or keel. The framework was made of willow and covered with sheeting fabric. The weight of the whole apparatus, without the operator, was 39.6 lbs.

In order to become accustomed to the management of these artificial wings, Herr *Lilienthal* first practised in his garden. Here he had a spring-board, toward which he ran for a distance of about 26 ft. ; and with the velocity thus acquired, together with the reaction of the spring-board, he launched himself into the air, where he could learn to operate and to manage the wings.

After these preliminary experiments had given him dexterity and facility in the management of the apparatus, he betook himself to a hilly region in the suburbs of Berlin, and there practised soaring flight in natural winds of moderate velocities. The plan, of course, consisted in first running against the wind, and thus deriving therefrom the necessary sustaining air pressure.

Having selected a hill whose downward inclination faced the prevailing wind, he ran along the summit straight toward the wind, until a sufficient velocity was attained at the brow, where he was carried into the air and landed safely at the foot of the hill, having sailed a distance of 65 to 82 ft.

When the wind velocity became greater than 11 to 13 miles per hour, the management of the apparatus became exceedingly difficult, and Herr *Lilienthal* advises an experimenter not to venture to leave the ground under such circumstances, unless he has attained, through long practice, a considerable degree of dexterity in manœuvre.

The results attained in the practice of the season of 1891 were sufficiently encouraging to warrant the further prosecution of these experiments in the following year; but they disclosed a number of points to which additional attention needed to be given in order to overcome the practical difficulties in imitating the birds. These points related to a better adjustment of the center of gravity, to methods for obtaining greater stability, and to the mode of management of the apparatus when the wind blew more rapidly than 11 to 13 miles per hour.

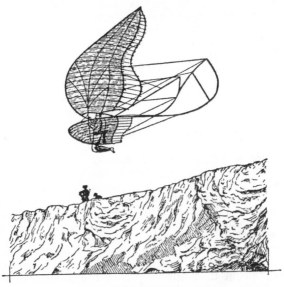

FIG. 74.—LILIENTHAL—1892.

In the issue of the *Zeitschrift für Luftschiffahrt* for November, 1892, Herr *Lilienthal* published an article on "Soaring and its Imitation," in which he gives a brief account of his experiments in the summer of 1892, from which the following abstract has been prepared:

Many theories have been proposed to explain soaring. My own explanation is based upon the advantageous relations of air resistances incident to the use of slightly curved wing surfaces (as I have demonstrated) and upon the gently rising trend of air current which I have found to prevail.

A flying apparatus which has the same proportions as those of a good soaring bird and is of sufficient size to carry a man, can scarcely be held fast by three or four men together when exposed to a brisk wind. When we look at the safe and quiet sailing of the birds, it almost seems as if some undiscovered mechanical principle were at work, some feature in the elastic properties of air or in the elastic curvature of the feathers which accounts for the mystery of sailing flight ; but my experiments have taught me that there is no mystery, and that the same mechanical science which has explained the theory of the steam-engine and followed the orbits of the planets is adequate to explaining the operations of soaring flight.

Dexterity alone, in my opinion, invests the native inhabitants of the air with superiority over man in that element. . . . Inasmuch as continuous soaring with large wings in high winds can terminate in scarcely anything but the destruction of the foolhardy fellow who may first attempt the experiment without previous practice, I first undertook last year to gain some expertness with a smaller apparatus and in moderate winds. In spite of my caution the wind several times played the mischief with me. Even with only 86 sq. ft. of sustaining surface, I was several times tossed up into the air by unexpected gusts of wind, and but for the circumstance that I was able to release myself quickly from my apparatus, I might have had a broken neck instead of the sprains in feet or arms which always healed in a few weeks.

Almost every Sunday, and sometimes on week days, I went out to practise on the hill between Grosskreutz and Werder. A mechanic, Herr Hugo Eulitz, the maker of my apparatus, went with me, and each practised alternately while the other rested. Thus we obtained dexterity in gliding down on the air and in landing at the foot of the hill without mishap.

Herr Kassner, of the Meteorological Institute, was so kind as to photograph me in the air, and has thus enabled me to exhibit to the members of the society how I sailed right over the head of the miller of Derwitz (in whose barn I stored my apparatus) and of his esteemed poodle dog.

Equipped with the experience gained in 1891, I this year attempted to soar with wings measuring 172 sq. ft. in surface. My apparatus weighed 53 lbs., and my own weight is 176 lbs., so that the whole was 229 lbs. Each square foot of surface, therefore, sustained $229 \div 172 = 1.33$ lbs.

The up-thrust of the wind (the lift) upon the wing surface is perhaps half as much as the pressure of the same wind upon the same surface if turned perpendicular thereto.* Now, as the apparatus therefore needs to sustain it a wind producing a pressure of $1.33 \times 2 = 2.66$ lbs. per square foot, we see that (by ordinary tables of wind pressures) it must blow at a velocity of about 23 miles per hour.

I have, however, been very cautious about exposing myself to such a wind with this large apparatus ; and in such high winds have used smaller surfaces for my sailing practice.

This year I selected a locality between Stegiitz and Südende. It had, however, the disadvantage that only westerly starts were possible. Herr Kassner has again taken instantaneous photographs of my apparatus, which have been laid before

---

* Lilienthal, " Der Vogelflug als Grundlage der Fliegekunst," Tafel VII.

the society (fig. 74). The strongest winds in which I practised had a velocity which I estimated at between 15 and 16 miles per hour. By running I obtained an additional velocity of 7 miles an hour, making the total relative velocity 23 miles an hour, which was required for soaring. Under these circumstances the first part of my flight was almost horizontal, and the alighting was always a gentle one. . . . Each apparatus had a vertical and horizontal tail, without which it is impracticable to practice in the wind. In conclusion, I will remark that sailing flight near the earth's surface must be much more difficult than at greater heights, where the wind blows more regularly, while every irregularity of the ground at lower levels starts whirls in the air.

In the opinion of the writer of these lines Herr *Lilienthal* has attacked the most difficult, and perhaps the most important, of the many problems which must be solved before success can be hoped for in navigating the air with flying machines. He has engaged in the effort to work out the maintenance of equilibrium in flight, and to learn the science of the bird. He has made a good beginning, and seems to be in a fair way to accomplish some success in riding on the wind.

We have already seen that this has been tried before, and that (to say nothing of ancient myths) *J. B. Dante, Paul Guidotti, Francisco Orujo*, and Captain *Le Bris*, all met with partial success in soaring. Singularly enough all four met also with the same accident—*i.e.*, a broken leg, in consequence of the loss of equipoise. Herr *Lilienthal* has greater chances of success, not only because he seems to have set about his experiments only after thorough investigation and consideration, but also because mechanical knowledge as well as constructive methods and workmanship have greatly improved since even *Le Bris's* time. Besides this, we have the gliding exploit of M. *Mouillàrd*, whose experiment has already been related, and that of M. *Ader*, which is yet to be mentioned.

Most of the capable inventors who have undertaken to solve the problem of flight have first concerned themselves with the question of motive power, and we shall see hereafter that very great progress has been achieved in this direction since 1890; but no amount of motive power will avail unless the apparatus to which it is applied is stable in the air—unless it can rise, sail, and come down again without danger of losing its equipoise. As has already been said, safety is the first requisite, and until this is assured, all the other elements of success will be unavailable.

Herr *Lilienthal* has eliminated for the present the question of motive power, by undertaking to utilize ascending trends of wind, like a sailing bird ; and if he succeeds in gliding up as well as down, and to the right or left, and in maintaining at all times the coincidence of the center of gravity with the center of pressure at all angles of incidence, he may not only apply an artificial power hereafter, for use when great speed is required or when there is no wind, but he will also probably have evolved a method of gratuitous transportation through the air when the wind blows under proper conditions ; for there seems to be no good reason why a soaring apparatus for one man should cost more than twice as much as a first-class bicycle, or half as much as a city carriage ; and when the wind is in the right direction, a good many miles could be sailed over in a day with no expenditure of force save for the evolutions necessary to maintain the equilibrium, although this can only be done under peculiar circumstances, and the commercial use must be very much less than that of bicycles.

That this expectation is not altogether absurd will appear from a brief consideration of the power of the wind ; and to make the matter plain we will suppose it to have an upward trend of 15° or 26 per cent. or a very moderate inclination, which must be frequently exceeded. Under that circumstance a horizontal aeroplane will, as previously explained, have the horizontal component of the normal pressure directed to the front and acting as a forward propelling force. We may now calculate what the effect of this would be upon Herr *Lilienthal's* aeroplane.

This was proportioned in the ratio of 0.75 sq. ft. of surface to the pound of weight ; but as the surfaces were concavo-convex, we may assume that the coefficient of efficiency would be about the same as that which we have assumed heretofore for the pigeon, or 1.3 per cent. of the actual surface, and we may further simplify the calculations by assuming the equivalent *plane* surface as equal to 1 sq. ft. per pound to be sustained. Now if this be exposed to a wind blowing at the rate of 25 miles per hour, at which the rectangular pressure, as given by Smeaton's table, is 3.125 lbs. per square foot, and if we suppose the plane to be inclined forward, so as to point 5° below the horizon, then the wind will make an angle of 10° with the plane, at which the normal pressure, by our tables, will be 0.337 of the rectangular pressure. As the effect upon the plane is in the ratio of the angle which the latter makes with the direction in which we desire to calculate it—*i.e.*, the horizon, and this angle is 5°, the *sine* of which is 0.087, then we have for the propelling force for each square foot of sustaining surface :

Drift = 1. × 3.125 × 0.337 × 0.087 = 0.0916 lbs per square foot.

But as the speed is 2,200 ft. per minute, we have for the power :

Power = 0.0916 × 2200 ÷ 33000 = 0.00611 horse power per square foot,

which for an apparatus with 172 sq. ft. of sustaining surface furnishes a motive power of

0.00611 × 172 = 1.05 horse power,

which is the power at the disposal of Herr *Lilienthal* when the wind blows 25 miles per hour, with an upward trend of 15°.

This, of course, varies with the trend and the strength of the wind ; but it will be noticed that with the data assumed it will amount to some 6 horse power for an aeroplane with 1,000 sq. ft. of sustaining surface—an amount which will probably be surprisingly great to those who have not considered the subject.

It will doubtless be objected that these calculations are all based upon the assumption that the wind has an ascending trend, and that this condition does not uniformly obtain, particularly at sailing heights 'above the earth, where the wind may be horizontal at the very time that experiment shows an ascending trend near the surface. This is granted ; it is acknowleged that the calculations of power to be obtained from the wind are predicated upon an assumption which may be untrue part of the time ; but the answer to the main objection is that *the birds soar at all times* when there is wind enough (not too much), and that while we cannot yet explain how they do it, man ought to be able to avail himself of the same circumstances as the birds, if only he can maintain his equilibrium.

This is what Herr *Lilienthal* has undertaken ; he has done so with great prudence and good sense, and so far as the results of his experiments have been published they teach several valuable lessons, which may be summed up as follows :

1. The upright position for the body of the aviator is the most favorable, as being most natural to man.

2. Safety while learning the management of an apparatus is promoted by beginning with comparatively small surfaces, because wind gusts are liable to destroy the balance. It is best to glide downward in initial experiments until practice has conferred the skill requisite to maintain the equilibrium, in case the apparatus is tossed up in the air by the wind. This is a lesson which was not obvious, and it should be heeded by experimenters, some of whom have assumed that safety was best promoted by large surfaces.

3. The aviator must be so affixed to his apparatus that he can detach himself instantly should the machine take a sheer.

4. It is not safe to experiment in winds blowing more than 23 miles per hour until skill has been acquired in the management of the apparatus, or until the latter has been so improved as to minimize the danger.

5. It seems now reasonably possible for designers of soaring machines (and the writer knows several) to experiment with their apparatus without further search for some hidden secret, for Herr *Lilienthal* says that his experiments have taught him that there is no mystery about sailing flight ; that the wind is sufficient to account for it. Inventors need not look for some new mysterious force, some "negative gravity," * like that in Mr. Stockton's tale, to take them up into the air ; nor need they be afraid that if they propose to experiment with soaring machines they will be considered lunatics. The main question for them to consider is that of the equilibrium.

Of course, even if this be worked out, the practical usefulness of a soaring machine would be very limited. It could only be availed of when the wind blew with about the favorable velocity (neither too slow nor too fast), and its field of daily use would probably be limited to the trade wind latitudes, or, in other words, to those regions inhabited by the sailing birds ; but if the equipoise be worked out, if man succeeds in devising an adequate soaring apparatus and in learning how to use it, unhampered by the necessity for looking after a motor at the same time, it will not probably be long before some motor is added to confer upon him command of space at all times.

In June, 1891, the quidnuncs in Paris were interested in the rumored success of some experiments with a flying machine carried on near Paris, in the private park of Mr. E. Pereire, the banker, by M. *Clement Ader*, who was

---

* One theorist expounds his ideas as follows : "One point I have studied, and that is, How can a twenty pound wild goose carry itself so easily ? Weigh every feather you can pick off from a wild goose, and they will not weigh one pound. Now if the feathers be picked off from the goose he can come no nearer flying than we can.

"So there we have it clearly demonstrated that one pound of goose feathers can pick up nineteen pounds of goose and carry this nineteen pounds and its one pound of feathers through space at about half a mile a minute, if in a a hurry.

"Now my theory is this, and it applies to all birds. Notice any bird when he suddenly starts to fly, and you will notice a lightning-like quiver of his feathers. I believe that this quiver causes the production of a negative force of magnetism, or some kind of force which pushes the bird from the earth—just the reverse of the loadstone. He then has only to use his wings to propel the body, for the magnetic negative earth-force does the lifting, and that is all produced by the feathers. If it were not, then the bird ought to fly when divested of his feathers. This is the force which should be looked for ; whoever discovers it will make a fortune."

said to have succeeded in rising to a height of about 60 ft., and in flying a distance variously estimated at 100 to 400 yds.

M. *Ader* is a well-known French electrician, the inventor of a telephone, and has long been interested in the flying-machine problem. In 1872 he constructed an artificial bird 26 ft. across and weighing some 53 lbs., with beating wings actuated by the muscular force of the operator's legs, aided by elastic auxiliary pectorals. In high winds, and restrained by ropes in order to guard against accidents, it would lift up a man, but it was found, as many times before, that man has not the requisite energy to sustain his weight in calm air. Subsequently the same apparatus, or a modification of it (for the accounts are not quite clear), was set up under a shed at Passy, and visited by M. *de la Landelle*,* who states that the operator was stretched horizontally (a bad position) between the wings, and worked with his feet and hands the organism of transmission to the parts that acted upon the air. A certain lifting effect was produced, but not enough to sustain the whole weight. This apparatus was never photographed, but its inventor now contemplates unboxing it and setting it up again as a curiosity.

In 1891, as already mentioned, M. *Ader* built another artificial bird 54 ft. across, with which he experimented in the open air with such close privacy as he could secure ; but the details are being kept secret, as the inventor states that he believes that it is destined to play an important part in the national defense of his country. He merely mentions the fact that the motor and the man who works it are placed in the interior of the machine, which is shaped like a huge bat ; that the motor is actuated by a " mixture of a combination of vapors," and that the instrument of propulsion is a screw (of which he tried some eight patterns) placed at the head ; that the whole apparatus rests upon skates or upon wheels, and that he needs a long smooth, flat space to gather headway by sliding or rolling some 20 or 30 yds. or more. He stated that he had already expended some $120,000 in his aerial experiments during the 15 years that he had been working at the problem, and that he contemplated exhibiting his machine in the air, if he could secure the use of the great machinery hall built for the Paris Exposition of 1889.

The above data are extracted from an account of an interview with M. *Ader*, published in the Paris *Temps* of July 9, 1891, in which he gave an interesting account of the preliminary studies that led to his last conception, the

* Dans les airs, G. de la Landelle, pp. 236, 237.

result, as he says, of a private theory of the resistances of air, which he proposes to publish some day.

Moved, probably, by the accounts of the sailing of large birds published by M. *Mouillard* as witnessed by him in Africa, M. *Ader* first obtained from the zoölogical gardens some eagles and some large bats, and observed their flight in his workshop. Judging this to be insufficient, he next went to Algeria, but could find none of the large vultures near Constantine ; so, disguising himself as an Arab, he went into the interior with two Arab guides, and by enticing the birds with pieces of meat left in secluded places, he succeeded in obtaining ample observations.

M. *Ader* states that he became fully convinced that these vultures, some of them measuring 10 ft. across, do not beat their wings when rising on the air ; that they flap them at most two or three times when first rising from the ground, and then hold them rigidly spread out to the current of wind upon which they ride, and upon which they rise in great circling sweeps by merely adjusting their aeroplane to the varying conditions of incidence and force of wind.

Starting from his theory and observations, M. *Ader* next built the machine which he has been experimenting with near Paris, in the presence, it is said, of only three or four persons, and with many precautions to avoid divulging his secret. He has even announced that he intends, from patriotic motives, to take no patents in foreign countries, so as not to divulge the design of his apparatus, and that all he can say at present is that the problem is an exceedingly difficult one, involving enormous mechanical difficulties, which increase rapidly with the size of the apparatus.

Naturally this reticence excited curiosity, and the French paper *L'Illustration*, in its issue of June 20, 1891, published a picture from which fig. 75 is reproduced, and it also made the following comments :

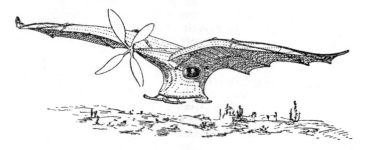

FIG. 75.—ADER—1891.

Nobody has seen anything, nobody knows anything, but *L'Illustration* has its friends everywhere. One of them was hunting lately in the environs of Paris, when he caught a glimpse through the leaves of a strange object resembling an enormous bird of bluish hue. It was impossible to approach close to it ; an enclosure surrounded the private park shut in by the forest in which the aforesaid machine was situated. Assuredly it could only be a flying machine. Our friend is something of a limner as well as an engineer, and he communicates to us the sketch which he made from a distance, and which is as correct as it was practicable to make it. Upon making due inquiry it turns out to be the invention of M. *Ader*, the electrician, well known for his telephone apparatus, and it seems that the machine has really flown several hundred yards, rising some 50 to 65 ft., and holding a course through space. . . .

The name of the inventor of this machine should be a guarantee of its possible success ; still we have our doubts. It is said to have glided a certain distance in the air—100 or 200, or, say, 400 yds But can it continue to do so for several hours, without having recourse to some fixed supply of power to recharge the motor actuating it? For this is the vital point : what is the motor? As the inventor is an eminent electrician, thoroughly understanding this new science, he must have selected his favorite motor, the dynamo.

But electric accumulators are impracticable on account of their weight, while primary batteries act for only a short time, and they, too, are heavy.

Therefore, for the present, and until we have witnessed a convincing experiment, at which we shall have seen with our own eyes the generator of the power employed, we shall remain skeptics, and we shall believe (and this only because of the high scientific standing of the inventor) that if the machine sketched by our friend can really fly, it is only for a very brief period of time.

In point of fact, it is surmised by the writer of these lines that M. *Ader* has really been experimenting with a soaring machine, using a motor only to get under way, and (if the sketch of the apparatus is correct) that the principal difficulty he has met with has been to maintain the equilibrium. He may have had a few good flights under favorable circumstances, but he must have had many mishaps.

It is probable that one of his errors lies in adopting too large a sustaining surface, under the mistaken belief that this would promote safety. It would probably do just the reverse, by enabling little wind gusts and ground currents to upset the equipoise. The machine is 54 ft. across, and must spread to the breeze twice the surface employed by Herr *Lilienthal*, which we have already seen is found by the latter to be dangerous in winds of more than 23 miles per hour.

In August, 1891, M. *Trouvé,* whose mechanical bird
with flapping wings actuated by explosions within a Bourdon tube, and whose hovering screw machine, worked by
a dynamo connected by a wire to a source of electrical
energy remaining on the ground, have already been noticed, deposited with the French Academy of Sciences a
sealed letter, containing descriptions and drawings of an
aeroplane, which he believes to be destined to solve successfully the problem of aerial navigation.

This method of depositing sealed descriptions of inchoate inventions with the Academy of Sciences is a favorite one in France, and answers generally much as the filing
of a caveat does in the United States.

Nothing is known, of course, concerning the designs for
this aeroplane, but M. *Trouvé* says that he has made great
strides toward developing his aerial apparatus since 1870,
and especially since 1884 ; that his laboratory experiments
have convinced him that while his explosion motor is satisfactory as to the power exerted in proportion to weight,
wings are less efficient than screws as instruments of propulsion. He has therefore designed an aeroplane propelled by two screws, rotating in contrary directions,
which he believes to be superior to the former arrangement of beating wings.

The arrangement of this aeroplane is said to be such
that the surface may always be proportioned to the weight
to be carried, no matter what that weight may be.

The method of obtaining initial velocity is ingenious and
effective. The apparatus is to be placed upon a railway
car, and this is to be towed by a locomotive upon an ordinary railway, until the speed is sufficient to furnish the
required reactive support from the air ; when, the machine
rises, and is thenceforth supported by its sustaining surfaces, driven by the two screws moved by the explosion
motor.

M. *Trouvé* believes that success is now a simple question of money expenditure, and that the daring man,
favored by fortune, who first navigates the air will reap
the glory of that success with less title thereto than his
predecessors, who have pointed out the way.

In 1891 *Gustav Koch,* an aeronaut of Munich, published
a pamphlet entitled " Free Human Flying, as the Preliminary Condition of Dynamic Aeronautics," * which
contains the plan and description of an apparatus designed
by him, in order to endeavor to imitate the soaring of
the birds, and which also gives an account of the experiments which he had tried with models. This design has

---

* Der freie Menschliche Flug, als Vorbedingung dynamischer Luftschiffahrt. München, 1891.

been thought worthy of trial, and the Bavarian Ministers of the Interior and of Education in May, 1893, granted 1,600 marks ($400) to Herr *Koch* to enable him to make experiments. This he is about to do (with an assistant) over the lake of Constance near Lindau, and while the results may not prove satisfactory, they cannot but prove interesting.

The aeroplane designed by Herr *Koch* consists in a pair of rigid wings, approximately shaped like those of the dragon fly, each about 27 ft. long and 6 ft. broad, back of which there is a triangular tail, some 7 ft. long and about 8 ft. wide at the rear end. The wings are to be constructed of bamboo, covered with unbleached silk slightly oiled, and pivoted to the back of the operator. The latter is to lie horizontally, face downward, in a sort of hammock suspended from a frame which attaches to the wings, and the latter can thus be swung forward or back within small limits, so as to change their position with respect to the center of gravity, but they have no flapping action whatever. The operator is to swing the wings and to elevate or depress the tail by means of pedals on which his feet rest, and of lines leading to his hands.

It will thus be seen that the action of the apparatus, which is some 57 ft. across, consists in altering the position of the center of pressure, with respect to the center of gravity, by swinging the wings forward or back, and thus changing the angle of incidence which the apparatus makes with the course, while still further changes can be produced by the action of the tail.

The weight of the aeroplane, including the mechanism which works it, is estimated at 99 lbs., and that of the operator at 176 lbs., making a total of 275 lbs., to be sustained by about 325 sq. ft. of surface.

Herr *Koch* proposes to test the apparatus by taking it up beneath a balloon and cutting it loose when about 3,000 ft. in the air. The first experiments, of course, are to be tried with a dummy instead of a man, and if these indicate sufficient strength and stability, the operator is to take the place of the dummy. He expects the machine to descend like a stone for the first second or two, and then, when air pressure has gathered under the wings, to gradually right itself, and to glide downward upon an easy slope, which would bring it down to the water in about 8 minutes and a distance of some 2½ miles, thus being a dirigible parachute. Meanwhile, however, the operator is expected to bring the apparatus under control ; by swinging the wings forward he expects to tilt the planes so as to glide upward again, by virtue of the acquired momentum, and by movements of the tail and of his own body, which has a certain latitude of motion in the hammock, he expects to tack and to sail upon the wind like a soaring bird, sweeping in circles or making a series of zigzag glances, during which elevation might be gained by utilizing the force of the wind.

Such is the scheme; it is not wholly devoid of merit, because the soaring birds perform those very manœuvres, and they do it much in the way which Herr *Koch* has indicated, but it may be questioned whether his apparatus is properly designed to accomplish the results desired. In the first place, the sustaining surface and the spread across are too great, and will terribly strain the strength of materials. It would be better to shorten the wings and to make them broader in order to reduce the length of leverage. In the second place, the horizontal position selected for the operator, probably to reduce horizontal resistance, is decidedly bad, because it is unnatural to man, and gives him inadequate control over the apparatus. The man should be placed vertically, and instead of manœvring, as planned, to cause the back part of the tail to strike the ground first and roll along, while the aeroplane settles forward slowly, the operator should alight on his feet and stop his impetus while running if he alights on land. In the third place, the mode of experimenting proposed is exceedingly dangerous. Herr *Koch* says, quite properly, that the first step toward success in artificial flight consists in acquiring the skill to manage an apparatus, but until that skill has been acquired it will evidently be little short of suicide to cut himself loose high in air, even if over a bed of water.

Perhaps, however, these various elements of failure have already been eliminated. The design was published in 1891, and may by this time have been so remodelled as to lead, not to an absolute success, for this is not to be expected, but to such partial control over the apparatus as to warrant further experiments.

In the *Cosmopolitan Magazine* for November, 1892, and in *Cassier's Magazine* for February, 1893, appeared two analytic articles by M. *J. P. Holland*, in which he takes the ground that mechanical flight has already been proved to be attainable, that what remains to be done is merely to combine things already tried and proved by other experimenters; and in which articles he advances three proposals or designs for flying machines.

In the *Cosmopolitan* article M. *Holland* proposes to place two aerial screws, superposed and rotating in contrary directions, above a spindle-shaped body containing the machinery, with a pair of wings or aeroplanes attached. This may be termed his first design, as indicated

by his figs. 2, 3 and 4. In his second design the spindle
and the superposed screws are retained, but the supporting surface consists of 10 narrow, superposed, concavo-
convex aeroplanes, somewhat like a Venetian blind, and
they as well as the screws are mounted upon a frame
pivoted to the spindle-shaped body, so that the screws
may first be used to raise the apparatus from the ground,
and then to drive it forward when the frame is raised to
the vertical ; support being then derived from the aeroplanes. This is indicated in M. *Holland's* figs. 5, 6 and 7.

In the *Cassier's Magazine* article the design is further
modified by placing the aerial screws side by side in the
frame instead of superposing them. The superposed aeroplanes are retained, but the number is increased to 16,
and the mode of operation is much the same.

The design is somewhat similar to that which Mr. *Phillips* experimented in England, which was illustrated in
Engineering of May 5, 1893, but is an improvement upon
the latter design in the provision for pivoting the Vene-
tian-blind aeroplanes to the body, and in the employment
of two screws instead of one for the propelling instrument.

If there be one man, more than another, who deserves
to succeed in flying through the air, that man is Mr.
Laurence Hargrave, of Sydney, New South Wales. He
has now constructed with his own hands no less than 18
flying machines of increasing size, all of which fly, and as
a result of his many experiments (of which an account is
about to be given) he now says, in a private letter to the
writer, that : " I know that success is dead sure to come."

M. *Hargrave* takes out no patents for any of his aerial
inventions, and he publishes from time to time full accounts of them, in order that a mutual interchange of
ideas may take place with other inventors working in the
same field, so as to expedite joint progress. He says :
" Workers must root out the idea that by keeping the results of their labors to themselves a fortune will be assured
to them. Patent fees are so much wasted money. The
flying machine of the future will not be born fully fledged
and capable of a flight for 1,000 miles or so. Like everything else it must be evolved gradually. The first difficulty is to get a thing that will fly at all. When this is
made, a full description should be published as an aid to
others. Excellence of design and workmanship will
always defy competition."

M. *Hargrave* is probably correct in his reasoning ; for
the history of all new methods of transportation teaches
that the original inventor seldom receives pecuniary re-
ward for the contrivance which is the first to succeed, but
nevertheless he is certainly broadly liberal in giving to

the world gratuitously the results of his constant studies
and labors. He uses exceeding care in determining the
different elements which compose the flight of his models.
He has carefully registered the sizes of all the parts, the
power consumed in each performance, and the length of
the flight, together with its trajectory. He states that he
has always kept his work in such shape that it could be
taken up and continued by any person at any time ; so
that a stranger, if an expert, could come into his shop,
study his notes and drawings, pick up his tools and
continue his work, and thus no portion of it would be
lost.

M. *Hargrave* reports regularly the progress of his work
to the Royal Society of New South Wales, of which he is
a member. Thus far 13 such papers have been published,
the latest having been read June 7, 1893.

He first devoted his attention to the motions performed
by the propelling surfaces of birds and fishes, the waves
which these created in the fluids on which they acted, and
the counteraction of these waves upon the forms of the
propelling surfaces themselves. The first paper, therefore, presented in August, 1884, was on the *Trochoided
plane*, which M. *Hargrave* defines as " a flat surface, the
center of which moves at a uniform speed in a circle, the
plane being kept normal to the surface of a trochoidal
wave, having a period equal to the time occupied by the
center of the plane in completing one revolution." This
was illustrated by working models, and the motions of
wings and of fishes in swimming were artificially reproduced.

Starting from these data, M. *Hargrave* next experimented with nearly 50 models intended to reproduce horizontal flight, and in exhibiting some of these and reading
his second paper, June, 1885, he said : " If the motion is
not that used by birds, it is at all events very like it, and
its acceptance or rejection as a scientific truth is of no
further interest, as it only remains for practical mechanics
to step in and adjust the details to suit the material and
the motive power which they may think best for the purpose they have in view ; or, in other words, that the solution of the problem of just how a bird flies is of very trifling
importance from a practical standpoint, as compared with
the judicious variations of the parts of the machine that
will have to be made before any return can he expected
for money invested in such undertakings."

Some of these models seem to have been driven by
clock-work, and the motions were those of the " trochoided
planes," as applied to flapping wings ; then selecting the
best of these models, and making their mean dimensions

a standard from which to take a fresh departure, M. *Hargrave* next built a series of experimental flying machines, actuated by india-rubber in tension.

The French experimenters, as we have seen, have preferred to use rubber in torsion in order to diminish the strains upon the central spine or backbone of the model, but they thus obtained less energy per pound of weight than if they had used it in tension. M. *Hargrave* stretched the rubber so that its elongation was multiplied by pulley-tackle, and that, as the rubber contracted, its center of gravity moved forward, thus advancing the center of gravity of the entire machine, and consequently diminishing the angle of flight as the force of the rubber decreased.

No less than 10 different flying machines of various types were thus built and experimented with, all moved by rubber in tension. In the first models the cord proceeding from the rubber was wound around a cylindrical drum on the crank-shaft, but owing to the variable resistance natural to a crank-shaft, it was found better to replace the cylindrical drum by a flat winder, so adjusted on the shaft that the moment of the cord varied with the resistance of the crank, and thus communicated a more uniform movement to the wings.

Seven of these machines seem to have been propelled by flapping wings—*i. e.*, "trochoided planes"—but in order that a comparison might be made, three varieties of models were made with screw propulsion—namely, with double and with single screws in the bow, and with a single screw in the stern, which latter was concluded to be the most practicable and serviceable form.

From these experiments M. *Hargrave* concluded that the screw and the flapping wings are about equally effective as instruments of propulsion, although he rather prefers the latter, as the wings possess several marked advantages. Any currents, he says, initiated during the up-stroke are utilized in giving increased efficiency to the down-stroke, if the machine has not progressed far enough to be acting upon entirely undisturbed air. Moreover, when steam-engines come to be used, there will be only one cylinder needed for both wings, there will be no conversion of reciprocating into rotary motion, and no variable listing moment to be counteracted, while, finally, there is less liability that wings shall be damaged in alighting than screw blades.

Fig. 76 shows the last one (1889) of the india-rubber driven machines described by M. *Hargrave*. He calls it the " 48 band-screw." The screw is at the stern, and the machine weighs exactly 2 lbs. Its sustaining area is 14.51 sq. ft. (7.26 sq. ft. per pound), and it flew 120 lineal feet with the expenditure of 196 foot-pounds of energy, while the preceding machine, weighing 2.09 lbs., with flapping wings, had flown 270 ft. with 470 foot-pounds, thus showing respectively 0.61 and 0.57 lineal feet flown per foot-pounds of power.

The framework of these machines was of pine, the larger piece (main spine) being a hollow box-girder, to secure strength and lightness. The sustaining surfaces were of paper, pasted on, and after the gum was dry rendered as tight as a drum by blowing a light spray of water over the paper and allowing it to dry. Thus with small, light, simple, and inexpensive models many experiments were made, and great advance realized in the distance flown over any previous experiments of others.

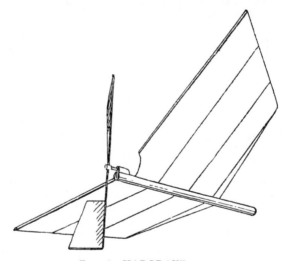

Fig. 76.—HARGRAVE—1889.

Having progressed thus far with india-rubber as a motive power, and gathered most valuable data and experience : as to the best arrangement and proportion of parts, the equipoise and the power required, M. *Hargrave* next undertook the construction of a flying machine actuated by compressed air, and, in 1890, he produced the machine illustrated by fig. 77, which he calls his '' No. 10, 40.5 oz. compressed air,'' and which marked a very considerable advance in design by a great simplification of the propelling arrangement.

In presenting it to the Royal Society, June 4, 1890, M. *Hargrave* said :

The principle embodied in this experiment is that of Borelli,

published in 1680, and it doubtless has had many stanch advocates in later times ; but the writer maintains that this is the
first practical demonstration that a machine can and does fly by
the simple (vertical) flapping of wings ; the feathering, tilting,
twisting, trochoiding, or whatever it may be called, being solely
effected by torsional stress on the wing arms.

The combination of Borelli's views with the results of work
recorded in your proceedings (Royal Society) has swept away
such a mass of tackle from the machine that its construction
becomes a ridiculously simple matter. The engine of the
model, of course, retains its precedence as the most important
part, and by continuous effort the number of pieces and the
difficulties of construction have been so reduced that it is possible to make them by the gross at a cost that cannot exceed
five shillings each ($1.25). For instance, the cylinder, usually
the most expensive portion of an engine, can be produced with
the ease and celerity of a tin can., . . .

It might be said that this flying machine is not on the principle
enunciated by Borelli, 'because the wings are not continuous
from their tip to the body. But this arrangement is only a device to enable the wing tips to act on the required quantity of
air with less spread ; it may possibly be one of those variations
which make all the difference between success and failure.
These wings are also distinctly double-acting, and it is not
quite clear that birds' wings thrust during the up-stroke ; but,
as previously stated, the question as to the exact movement of
a bird's wing is merely straw-splitting, when we have a mechanism that actually flies and is manifestly imperfect in its present mechanical details.

This machine flew 368 ft., with the expenditure (as corrected by M. *Hargrave*) of 870 foot-pounds of energy. It
weighed 2.53 lbs., and the sustaining body plane measured
14.78 sq. ft., while the two wings measured 1.50 sq. ft. in
area, making a total of 16.28 sq. ft., or, say, 6 43 sq. ft.
per pound.

London *Engineering*, in its issue of December 26, 1890,
gives the following description of the machine :

The compressed air is stored in a tube which forms the backbone of the whole construction. This tube is 2 in. in diameter,
48¼ in. long, and has a capacity of 144.6 cub. in. Its weight is
19.5 oz., and the working pressure is 230 lbs. per square inch.
The engine cylinder has a diameter of 1½ in. and a stroke of
1¼ in., while the total weight of the engine is only 6½ oz. The
piston-rod is made fast to the end of the backbone, and the
cylinder moves up and down over the piston. Two links connect the cylinder to the Canadian red pine rods which carry the
wings. The air is admitted to the cylinder and exhausted by
means of a valve worked by tappets. The period of admission
continues through the entire stroke. The cylinder and receiver
ends are pressed, and the piston is made of vulcanite, with a
leather cup ring for packing.

The wings are made of paper, and have no canting or feathering motion other than that due to the springing of the material of which they are made. The weight of the wings is 3 oz. To find how much the wings deflected, one was held by the butt and a weight of 7½ oz. was put on the membrane 24 in. from the fixed point, and 1⅝ in. abaft the wing arm. The deflection produced, due to torsional stress, was 3½°. By moving the weight half way across the wing it was twisted 8¼°. The area of the body is 2.128 sq in.; the area of the wings 216 sq. in., and the total area 2.344 sq. in.

When first made, the machine had its center of gravity so placed that the percentage of area in advance of it was 30 per cent. of the whole area, but continued disaster caused its reduction to 23.3 per cent. In a dead calm the machine flew 368 ft. horizontally.

It will be noted that the engine is a marvel of simplicity and lightness. Its cylinder is made like a common tin can, the cylinder covers are cut from sheet tin and pressed into

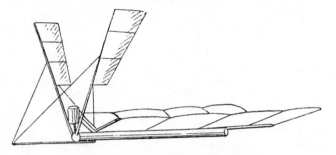

FIG. 77.—HARGRAVE, No. 10—1890.

shape in a vice, the piston and junk-ring are made of vulcanite, and the cup leather packing does away with the necessity for the cylinder being either round or parallel.

Beside the engine, a marked advance consisted in securing the torsion of the wings through no special mechanism, as formerly, but simply by the elasticity of the material composing them. This throws a new light upon the part performed in the flight of birds by the elasticity of their feathers, and promises great simplicity and efficiency in the future designing of artificial wings.

By looking at the figure, a bowsprit will be noticed. This was a so-called safety stick, which was added to break the fall of the machine when alighting, and it has proved quite successful in accomplishing that object.

A noticeable feature of this and subsequent machines exhibited by M. *Hargrave* consists in the extraordinary length of its supporting body plane. The same surface would carry a far greater load if it were driven broadside instead of lengthwise ; but M. *Hargrave* explains that the plane was purposely so designed in order to insure longitudinal stability. This quality might also be secured by placing a tail far in the rear of a narrow supporting plane, as practiced by *Pénaud* and others. He states, moreover, that the plane is rendered more effective per unit of surface by being cut away in the middle portion, or by being formed in two parts, separated by a gap.

As regards the lateral equilibrium, he seems to have met with but little difficulty ; a slight diedral angle of the two halves of the body plane with each other providing the necessary stability, and preventing any swerving, so long as the center of gravity was at all below the center of effort ; but he had great trouble in working out the longitudinal stability. This he did upon the " cut and try" principle—a method doubtless the most thorough, the surest, and the most convincing, but also the most tedious. He found that the direction up or down of the machines in flight was entirely due to the distance of the center of gravity from the forward edge of the body plane, and therefore to the coincidence or otherwise of the center of gravity with the center of pressure. He measured the percentage of area in advance of the center of gravity in his three most successful machines, and found it respectively 19 3, 20 and 23 3 per cent. of the length of the plane, while subsequently he came to the general conclusion that the true position for the center of gravity for a continuous rectangular surface is situated between o.25 and o.2 of the length from the forward end, these positions being arrived at " by experience gained by repeated wrecks when groping in comparative darkness."

This independent working out of a complex question well illustrates the perseverance and ingenuity of this experimenter. At this juncture, however, he would have been saved much groping, time, and annoyance had he been aware of the formula of Joessël for determining the center of pressure :

$$C = (\text{o.2} + \text{o.3 sin. } a)\ L,$$

in which *C* is the distance from the forward edge of a rectangular plane to its center of pressure, when inclined at the angle of incidence *a* with the course, and *L* is the length of the plane along the line of motion.

In the same year (1890) M. *Hargrave* built another flying machine, actuated by compressed air and propelled by beating wings. This is shown by fig. 78. It was of the increased weight of 4.63 lbs., with sustaining body plane of different shape, measuring 29.63 sq. ft., or in the proportion of 6.40 sq ft. per pound. It flew 343 ft., with an expenditure of 789 foot-pounds of energy, and therefore showed better results than the previous machine (No. 10), inasmuch as more pounds were transported on the air approximately the same distance, with a somewhat smaller expenditure of energy.

Having apparently found some advantage by shortening the body plane, M. *Hargrave* next built his flying machine No. 13, which s shown in fig. 79, with a body plane still shorter, and he provided it with a two-bladed aerial screw, set in the bow and actuated by a three-cylinder compressed-air engine of the Brotherhood type. This drove it 128 ft. in eight seconds, with an expenditure of 143 foot-pounds of energy. The apparatus weighed 46.86 oz. (2.93 lbs.), and exposed 2,952 sq. in. or 20.5 ft. of floating surface, being in the ratio of 7.00 sq. ft. per pound.

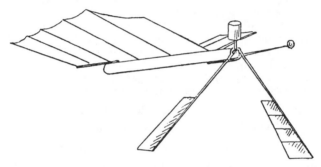

FIG. 78.—HARGRAVE, No. 12—1890.

This is the first time (paper 10, July 1, 1891) that M. *Hargrave* gives us the time of flight of his machines, so that we may calculate the number of pounds of weight transported in ratio to the horse power. He says :

The time of flight is taken with a sandglass which has a loop of string at each end of it. The loop at the sand end is put round the right wrist, and the other loop is held between the right thumb and the receiver, so that the glass is turned the moment that the machine is let go. On the machine taking the ground the glass is put horizontal, and the sand which has fallen is timed at leisure. This seems an obvious enough method of finding the speed, but a practical way to do it was not devised previously.

This showed for No. 13 machine a speed of 10.34 miles per hour, which is about what we should have expected from the large proportional surface, it being about in the ratio of the slowest flying birds. This low speed M. *Har-*

grave adopts on purpose, the better to observe the motions of the machines and to save breakage, and he adds quaintly that he sees no objection to this course, so long as the atmosphere is not crowded with flying machines. As No. 13 machine (fig. 79) is reported as having expended 143 foot-pounds in eight seconds, we have:

$$\text{Power} = 143 \div 8 = 18 \text{ foot-pounds per second,}$$

nearly, and, as it weighed (as reported) 2.93 lbs., we have for the weight sustained per horse power:

$$2.93 \times 550 \div 18 = 89.53 \text{ lbs. per horse power;}$$

while it will be recollected that M. *Tatin* sustained 110 lbs. per horse power and that M. *Phillips* in his recent (1893) experiments floated 72 lbs. per horse power. We will see by the analysis of subsequent performances that M. *Hargrave* did not obtain quite as good results with subsequent flying machines.

He next built his No. 14 flying machine, with much the same shape of body surface, but propelled by beating wings instead of a screw. It weighed 3.69 lbs. and exposed 22.84 sq. ft. of surface, being in the proportion of 6.19 sq. ft. per pound. It flew 312 ft. in 19 seconds, with an expenditure of 509 foot-pounds, and thus we have:

$$\text{Power} = 509 \div 19 = 26.79 \text{ foot-pounds per second,}$$

and for the weight floated per horse power:

$$3.69 \times 550 \div 26.79 = 75.75 \text{ lbs. per horse power.}$$

This apparatus (No. 14) M. *Hargrave* has generously offered to present to some American institution which will take proper care of it, believing it to be one in which "the increased skill in construction acquired by practice is thought to have resulted in an apparatus that, for its weight, it will be hard to excel." He says in his paper to the Royal Society:

It may be said that it is a waste of time to make machines of such small capabilities, and that no practical good can come of them. But we must not try too much at first; we must remember that all our inventions are but developments of crude ideas; that a commercially successful result in a practically unexplored field cannot possibly be got without an enormous amount of unremunerative work. It is the piled-up and recorded experience of many busy brains that has produced the luxurious travelling conveniences of to-day, which in no way astonish us, and there is no good reason for supposing that we shall always be content to keep on the agitated surface of the sea and air, when it is possible to travel in a superior plane, unimpeded by frictional disturbances.

No 16 was another compressed-air flying machine with beating wings and somewhat differently shaped body plane. It weighed 4.66 lbs., spread 26.06 sq. ft. of surface, and flew 343 ft. in 23 seconds, with an expenditure of 742 foot-pounds. The power was therefore :

Power $= 742 \div 23 = 32.26$ foot pounds per second,

and the weight floated per horse power :

$4.66 \times 550 \div 32.26 = 79.45$ lbs. per horse power.

Several forms of body plane seem to have been tested in this machine and no less than 12 trials were recorded, trial No. 10 being the successful one, from which the above data have been taken.

Having now constructed 10 flying machines of different types and proportions actuated by india-rubber in tension, and six actuated by compressed air, of increasing size and weight, M. *Hargrave* then turned his attention to produc-

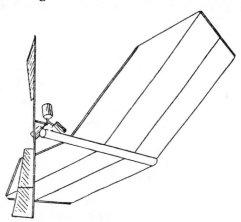

FIG. 79.—HARGRAVE, No. 13—1891.

ing a steam motor which should equal in lightness and surpass in power the best compressed-air motors thus far constructed by him, and which should furnish driving power for a longer time.

But first he endeavored to work out an idea which he seems to have entertained for some years, of testing an explosion motor. His engine No. 15 consisted of a turbine to be worked by the gases resulting from the explosion of a mixture of nitrate of ammonia, charcoal, and sulphur ; but a considerable expenditure of time only resulted in a failure.

He also experimented upon a method of utilizing sea

waves in propelling vessels, which he believes to be the germ of the solution of the soaring problem, and he succeeded in securing such automatic action that a 12½ lb. model advanced in the wind's eye at five-eighths of a mile per hour.

He also made some experiments upon pure aluminium, but found that it presented no advantages for flying-machine construction.

No. 17 flying machine of M. *Hargrave* is described in his twelfth communication to the Royal Society of New South Wales, read August 3, 1892. The total weight of the apparatus is 64.5 oz., or 4.03 lbs., including 12¾ oz. for the strut and body plane, so that the engine and boiler, including 5 oz. for spirit fuel and water, weighs 3.25 lbs., and develops 0.169 horse power, or at the rate of 1 H. P. per 19.2 lbs.—a very remarkable achievement.

The boiler is of the "Serpollet" type, made of 12 lineal feet of ¼ in. copper tubing (steel pipe could not be got in Sydney), in the form of a double-stranded coil, encased in asbestos, and placed just over the backbone of the apparatus. The fuel is methylated spirits of wine, drawn from a tank placed above the boiler, vaporized, mixed with air and spurted into the furnace. As much as 6.9 cub. in. of water have been evaporated by 1.7 cub. in. of spirit in 80 seconds, making 182 double vibrations of the propelling wings, say, 2.35 per second, and developing 0.169 horse power.

It was estimated that if the apparatus were loaded with 10 oz. more of spirit and water, and thus made to weigh the same as the compressed-air machine No. 12, which flew 343 ft., then the steam apparatus No. 17 would possess a sufficient store of energy to fly 1,640 yds., or nearly 1 mile.

But M. *Hargrave* has done still better, for in March, 1893, he prepared a paper, which was presented to the Conference on Aerial Navigation at Chicago, August 2, 1893, in which he gave data concerning his No. 18 flying machine. This apparatus is also driven by a steam-engine which weighs, with 21 oz. of fuel and water, an aggregate of 7 lbs., and indicates 0.653 horse power, or at the rate of 10.7 lbs. per horse power ; so that, roughly speaking, the weight of the motor has been doubled, and the power has been increased fourfold.

Four boilers were constructed. The final one was made of 21 lineal feet of ¼ in. copper pipe, with an internal diameter of 0.18 in., and arranged in three concentric vertical coils whose diameters were 1.6 in., 2.6 in., and 3.6 in. respectively. It weighed 37 oz., but it is now known " that a coil of equal capacity can be made weigh-

ing only 8 oz., and still excessively strong." The cylinder
is 2 in. diameter, with a stroke of 2.52 in. The feed-pump
ram is 0.266 in. diameter, and the piston valves 0.3 in.
diameter. On one occasion this motor evaporated 14.7
cub. in. of water with 4.13 cub. in. of spirit in 40 seconds.
During a portion of the time it was working at a speed of
171 double vibrations per minute.

M. *Hargrave* gives no data concerning the flight of his
last two (steam) machines. He states that 11 different
burners have been tried, and that the flame striking the
water boiler first has a tendency to vary the supply of heat
to the spirit holder. From this it is inferred that he is

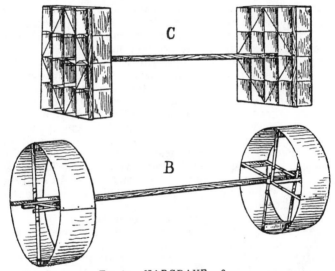

Fig. 80.—HARGRAVE—1893.

struggling with the same difficulties already encountered
by *Stringfellow*, by *Moy*, and by *Maxim* in regulating
and keeping alight spirit burners when the apparatus gets
under forward head-way ; but this difficulty, while a serious one, will doubtless be eventually overcome by persis-
tent experiment, and we may then expect flights of astonishing lengths.

Seeing now his way to an adequate motor and to extensive flights in the near future, M. *Hargrave* recently
turned his attention to experiments upon curved surfaces,
and to the seeking for a better disposition of the sustaining surfaces or body planes. He had described the eccen-
tricities of a curved strip in the form of a segment of a
hollow cylinder, when exposed to the wind, in his paper

No. 12 to the Royal Society of New South Wales, read
August 3, 1892, and he describes some of his experiments
with "cellular kites," in his paper read in the Aerial
Navigation at Chicago, August 2, 1893.

The "cellular kites" constitute quite a new departure,
and practically consist of superposed aeroplanes connected together in pairs. *B*, in fig. 80, shows the simplest
form. This consisted of two hollow cylinders of aluminium, each 13 in. diameter by 4½ in. deep, mounted 30 in.
apart upon a connecting stick, and weighing 14¾ lbs. The
kite-string was attached 11 in. back from the forward section, and as a consequence of the angle of incidence thus
produced, the apparatus mounted upon the wind. Its
particular behavior is not described in the paper. *C*, in
fig. 80, shows a kite with 16 cells, the length of each being
3 in., by a height of 3 in., and a breadth of 3 in. It was
made of cardboard, and the two sections were 22 in. apart,
the point of attachment of the kite-string being 6½ in. distant from the forward section, while the weight was 10.5
lbs. This seems to indicate that this kite flew at a steeper
angle than the preceding, although we should expect the
reverse, in consequence of the greater proportion of sustaining surface. M. *Hargrave* says, " These kites have
a fine angle of incidence, so that they correspond with the
flying machines they are meant to represent, and differ
from the kites of our youth, which we recollect floating at
an angle of about 45°, in which position the lift and the
drift are about equal. The fine angle makes the lift
largely exceed the drift, and brings the kite so that the
upper part of the string is nearly vertical."

Kites *E* and *F*, fig. 81, are of exactly the same size and
weight, consisting of one cell, 4 in. long, 10.7 in. broad
by 6.25 in. high, constructed of wood and paper, and
weighing 3.25 lbs.; the two sections are 21.25 in. apart,
and the string is fastened 7.25 in. back of the forward section. The only difference is that kite *E* has its horizontal
(top and bottom) surfaces curved to a radius of 4.5 in.,
while all the surfaces of kite *F* are true planes. The result is that when kite *E* is flown with the convex sides up,
it pulls about twice as hard on the string as kite *F*, so
that, as M. *Hargrave* says : " A flying machine with
curved surfaces would be better than one with a flat body
plane, if the form could be made with the same weight of
material."

M. *Hargrave*, in this last paper, figures and describes
two other forms of cellular kites with which he has experimented, and points out that the rectangular form of cell
is collapsible when one diagonal tie is disconnected, so as
to make it easy of transportation. He says : " Theoreti-

cally, if the kite is perfect in construction and the wind steady, the string could be attached infinitely near the center of the connecting stick, and the kite would fly very near the zenith. It is obvious that any number of kites may be strung together on the same line, and that there is no limit to the weight that may be buoyed up in a breeze by means of light and handy tackle. The next step is clear enough—namely, that a flying machine with acres of surface can be safely got under way, or anchored and hauled to the ground by means of the string of kites."

He duly gives credit to M. *Wenham* for suggesting the superposition of planes in 1866, and it is an interesting circumstance to note that at the same Chicago Conference,

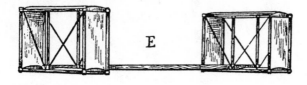

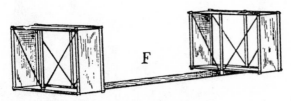

FIG. 81.—HARGRAVE—1893.

a paper from M. *Wenham* was read suggesting a course of experiments with kites, to determine the best arrangement of superposed aeroplanes and the conditions of equipoise.

Such are the labors of M. *Hargrave* up to the present time. He no longer troubles himself about the general problem of man's eventual success in navigating the air, but he says: "The people of Sydney who can speak of my work without a smile are very scarce ; it is doubtless the same with American workers. I know that success is dead sure to come, and therefore do not waste time and words in trying to convince unbelievers."

FIG. 82—MAXIM—1892

Instead of this, he constructs machines and reports the results in detail, so that others may repeat his experiments. He says that the record of unsuccessful experiments takes up a considerable portion of his notes, and further, that "there is no use in the mind's conceiving an idea, if the hands are not ready to carry out the work skillfully, in the absence of reliable assistance, and if the design be found faulty, the whole thing should be begun again without trying to use up old machines. The question of intricate workmanship and costliness is being continually battled with ; my constant endeavors are directed to making the machines simple and cheap, so that any one who doubts can verify my work, provided his hands are as skillful as mine, and I am sure that the photographs show clearly that the workmanship is anything but first-rate."

He began with small, cheap models, and has gradually enlarged their size, and obtained flights longer than any heretofore accomplished. It is noticeable that the heavier the model, and the smaller the sustaining area in proportion to the weight, the more successful has been the flight. He may not be the first man to ride at will upon the air, but he deserves to succeed.

In November, 1890, M. *Hiram S. Maxim*, the celebrated American inventor of a writing telegraph, of several systems of electric lighting, and of the " Maxim automatic machine gun, addressed a letter to the New York *Times*, in which he stated that, before sailing back to England, he thought it would be well to state what he was doing toward constructing a flying machine which had been alluded to lately by the American press. Among other things he said :

I would say that among the large number of societies to which I belong in England, the Aeronautical Society is one, and need I say that I am the most active member ? At the present moment experiments are being conducted by me at Baldwin's Park, Bexley, Kent, England, with a view of finding out exactly what the supporting power of a plane is when driven through the air at a slight angle from the horizontal. For this purpose I constructed a very elaborate apparatus, provided with a great number of instruments, and arranged in such a manner that I can ascertain accurately the efficiency of a screw working in air, the amount of power required to drive a screw, the amount of push developed by a screw, the amount of slip, and also the power required for propelling planes through the air when placed at different angles, as well as to ascertain the friction and all other phenomena connected with the subject. I have been experimenting with motors and have succeeded in making them so that they will develop 1 horse power for every 6 lbs. My experiments show that as much as 133 lbs. may be sustained in the air by the expenditure of

1 horse power ; of course, it is premature now to express any opinion ; still, if I am not very much mistaken, and if some new phenomenon, which I do not understand, does not prevent it, I think I stand a fair chance of solving the problem, and I think I can assert that within a very few years some one—if not myself, somebody else—will have made a machine which can be guided through the air, will travel with considerable velocity, and will be sufficiently under control to be used for military purposes. I have found in my experiments that it is necessary to have a speed of at least 30 miles per hour, that 50 miles is still more favorable, and that 100 miles would seem to be attainable. Everything seems to be in favor of high speed.

Whether I succeed or not, the results of my experiments will be published, and as I am the only man who has ever tried the experiments in a thorough manner with delicate and accurate apparatus, the data which I shall be able to furnish will be of much greater value to experimenters hereafter than all that has ever been published before.

In May, 1891, M. *Maxim* again visited the United States, and he gave to various newspaper reporters, notably to one from the New York *Sun*, some particulars concerning the flying machine, or "first kite of war," which he was building in England, and upon which he had spent up to that time (including the preliminary experiments) some $45,000.

He described the apparatus with which he had made his preliminary experiments, to ascertain accurately the supporting power and resistance of air to aeroplanes at small angles of incidence, and then continued as follows :

My large apparatus is provided with a plane 110 ft. long and 40 ft. wide, made of a frame of steel tubes covered with silk. Other smaller planes attached to this make up a surface of 5,500 sq. ft. There is one great central plane, and to this are hinged various other planes, very much smaller, which are used for keeping the equilibrium correct, and for keeping the flying machine at a fixed angle in the air. The whole apparatus, including the steering gear, is 145 ft. long. . . . A part of the aeroplane, or actual kite, is made of very thin metal, and serves as a very efficient condenser for the steam. . . .

It is ready and awaiting my return. It is now resting on a track 8 ft. wide and half a mile long, in my park. The first quarter of a mile of the track is double—that is to say, the upper track is 3 in. above the lower. By that means I am able to observe and measure the lift of the machine when it starts, because the upper track will hold it down when it lifts off the lower one. When completed the machine will weigh, with water tanks and fuel, somewhere between 5,000 lbs. and 6,000 lbs., and the power at my disposal will be 300 horse power in case I wish to use it ; but it is expected that about 40 horse power will suffice after the machine has once been started, and that the consumption of fuel will be from 40 lbs. to 50 lbs. per hour. The machine is made with its present great length so as to give a man time to think ; its length makes it easier to steer and to change its angle in the air. Its quantity of power is so enormously great in proportion to its weight that it will quickly get its speed. It will rise in the air like a sea-gull if the engine be run at full speed while the machine is held fast to the track, and if it is then suddenly loosened and let go.

M. *Maxim* very judiciously refrained from furnishing drawings or detailed descriptions of an apparatus which was still in process of evolution, and which he might want to modify as he proceeded in erection and trial. Indeed, it is probable that he has varied considerably from the various arrangements which he has patented from time to time,* so that drawings and descriptions made from these might be wide of the mark.

The important, the vital feature, however, he recognized to be the motor, and to perfecting this he gave his first attention. In steam motors he seems to have accomplished wonderful results, hitherto quite unreached, and in an article published in the *Century Magazine* for October, 1891, after describing and illustrating the experimental whirling machine with which he had gathered his preliminary data, he gives the following account of what he had accomplished up to that time with the motor :

I have come to the conclusion that the greatest amount of force with the minimum amount of weight can be obtained from a high-pressure compound steam engine, using steam at a pressure of from 200 lbs. to 350 lbs. to the square inch, and lately I have constructed two such engines, each weighing 300 lbs. These engines, when working under a pressure of 200 lbs. to the square inch, and with a piston speed of only 400 ft. per minute, develop in useful effect in push of screws over 100 horse power, the push of the screws collectively being over 1,000 lbs. By increasing the number of turns, and also the steam pressure, I believe it will be possible to obtain from 200 horse power to 300 horse power from the same engines, and with a piston speed no greater than 850 ft. per minute.† These engines are made throughout of tempered steel, and are of great strength and lightness. The new feature about my motors, however, is the manner of generating steam. The steam generator itself, without the casing about it, weighs only 350 lbs. ; the engine, generator, casing, pumps, cranks, screw-shaft, and screws weigh 1,800 lbs., and the rest of the machine as much more. With a supply of fuel, water, and three men, the weight will not be far from 5,000 lbs. As the foregoing experiments have shown that the load may be 14 times the push of the screw, it would appear that this machine ought to carry a burden, including its own weight, of 14,000 lbs., thus leaving a margin of 9,000 lbs, provided that the steam

---

* British patents Nos. 10,359 and 16,883, A.D. 1889 ; No. 19.228, A.D. 1891.
† The piston speed of an express locomotive is about 1,000 ft. per minute.

pressure is maintained at 200 lbs. to the square inch. The steam generator is self-regulating, has 48,000 brazed joints, and is heated by 45,000 gas jets, gas being made by a simple process from petroleum. When the machine is finished the exhaust steam will be condensed by an atmospheric condenser, made of a great number of very thin metallic tubes, arranged in such a manner that they form a considerable portion of the lifting surface of the aeroplane. The greater part of the machine is constructed from thin steel tubes. I found that these were much more suitable for the purpose than the much-talked-of aluminium ; still I believe that if I should succeed in constructing a successful machine, it would lead to such improvements in the manufacture of aluminium products that it will be possible to reduce greatly the weight of the machine.

The question of keeping the machine on an " even keel," of steering, and of landing, has been duly considered and provided for, but a description of these would be premature before the machine has actually been tried.

When it is remembered that locomotives weigh some 200 lbs. per horse power, that the lightest marine (launch) engines in 1889 weighed about 60 lbs. per horse power, and that the largest steam-engines previously built for aerial navigation purposes were those of *Giffard* and of *Moy*, each of 3 horse power and weighing (with their boilers) 110 lbs. and 27 lbs. per horse power respectively, then the importance of M. *Maxim's* achievement, as above set forth, may be partially realized ; particularly when it is considered that the relative weight tends to increase with the size, and that M. *Maxim's* expectations of obtaining 300 horse power from the same engines have been fully confirmed, as will be seen hereafter.

Moreover, as exhausting the steam into the air would involve carrying a supply of water amounting to some 20 or 25 lbs. per horse power per hour, and this would have been simply prohibitory, M. *Maxim's* plans included a surface aero-condenser, in order that the same water might be used over and over again. This was a wholly unsolved problem, such tentative experiments as had been tried previously by others having indicated weights of 50 lbs. to 150 lbs. per horse power, as necessary for efficient aero-condensers, and this would also have been prohibitory.

M. *Maxim* proposes to solve this problem by making all the frames of his apparatus of hollow tubes, and connecting therewith a condenser consisting of a large number of wide, flat, or film tubes—that is to say, of tubes of thin metal having a flat bore, through which the steam will pass in thin films of considerable width ; these film tubes being so arranged that in the forward motion of the machine the air will impinge upon them, thus effectually cooling them and condensing the steam therein. This aero-condenser is utilized as a part or the whole of the sustaining surface, or there may be substituted therefor a large flexible bag or chamber, connected at the forward part with the exhaust steam-pipe, and at the rear end with the hot well, or directly with the suction pipe of the feed-pump. He relies, of course, upon the increased condensation produced by air currents due to the forward motion of the machine, and the extent of the condenser is therefore a matter for experiment, so that its exact weight cannot be settled in advance.

The horizontal angle of incidence in flight is to be maintained by a " Gyrostat," which consists in a gyroscopic wheel rotating rapidly, suspended by universal joints and connected with two horizontal rudders, one at the front and the other at the back of the apparatus, so as to act upon them instantly (through the well-known property of the gyroscope to continue rotating in the same plane), in case any tendency occurs to deviate from the angle of incidence with the horizon.

The whole of the apparatus is to be thoroughly stayed by diagonal wire ties, so as to make every part rigid and prevent deformations under varying wind pressures.

Fig. 82, engraved from a photograph kindly furnished by M. *Maxim*, exhibits the main features of the apparatus. It does not show the front or back rudders, which have been removed, nor the side wings, set at a diedral angle, to preserve the transverse stability, nor sundry possible keel-cloths or auxiliary planes intended to promote the same object. It exhibits the central or principal aeroplane, with the forward end facing the observer. This main aeroplane is understood to be 50 ft. wide, about 58 ft. long, and slightly concave in the direction of its length, while it is trussed and stiffened in every direction by wire stays. The condenser is indicated by the dark shading at the front of the main plane, and, as will readily be seen, can be largely increased in surface, but, however, at the expense of added weight. The driving screws are placed at the rear, and are understood to be 17 ft. 10 in. in diameter, the speed of rotation varying, of course, with the power exerted.

The whole apparatus is mounted upon wheels, running over a railway track, so as to acquire sufficient speed to rise upon the air, and the three men who are grouped about the front may enable the reader to gather by comparison some general conception of the colossal dimensions of this flying machine.

Having thus designed and built his apparatus, the next point for M. *Maxim* to consider was how to get it up into

the air, how to control it while sailing, and how to alight
with it safely. To this he has evidently given much
thought, and in an article published by him in the *Cosmopolitan Magazine* for June, 1892, he thus describes what
course he would pursue if a sum of $100 000 were placed
at his disposal, for constructing and experimenting a successful flying machine ; which course seems to be so carefully planned that we may fairly assume that it is the one
determined upon by M. *Maxim* for experiments with his
own actual machine.

" The machine should be run around the one-mile track
at all speeds, from 20 miles per hour to 100 miles per
hour, and the power actually required should be carefully
noted.   These runs would enable us to ascertain how our
pumps worked at high speed, and how much our screws
pushed, and if we put a brake to the wheels we should
find out the slip of the screws.   We could also ascertain
the efficiency of our condenser at various speeds, and the
temperature of the water could be taken.   In order to run
on a railway track, the machine, of course, must be provided with wheels, and two sets of these would be necessary ; one set should be of great weight, so as to hold the
machine down when running on the track, and the other
set should be light, for actual flying.   Springs should be
interposed between the axletrees and the machine, after
the manner of railway carriages, and there should be attached above each wheel some sort of an index or indicator to show the exact load resting on each wheel.   When
all the parts of the machine had been made to operate
smoothly and satisfactorily, the silk could be placed on the
aeroplanes, and then our serious experiments might be
said to commence.

" We should first begin by running slowly —say at the rate
of 20 miles per hour—and carefully note the lift on the in-
dexes over each wheel.   If we found that with a speed of
20 miles an hour, three fourths of the load was lifted off
the forward axletree, and only one-fourth off the hind one,
then we should change the center of weight further forward, so as to bring it as near as possible under the center of effort or lift.   We should then make another trial,
and if we found that the lift was equal both fore and aft,
we should increase the speed very carefully, gradually observing the lift at the four corners of the machine, until
the whole weight of the machine was supported by the
aeroplane, and the whole weight of the wheels (about one
ton) by the railway track.   Then, when there was neither
lift nor load on either wheel, we might consider that we
had arrived at a stage in our experiments where we could
turn our attention to the subject of steering.

" A boat has to be steered in only one direction—namely, a horizontal direction, to the right or to the left. A locomotive torpedo or a flying machine must be steered in two directions—right or left, or up or down. We should experiment with the more difficult one at first—namely, the up and down or vertical direction. We should attach two long arms to our aeroplane in such a manner that they would project a considerable distance in the rear of the machine. To these arms we should pivot a very large and light silk-covered rudder and connect it with ropes, so that it could be turned up or down by a small windlass from the machine. We should then take a run on the track and see if the changing the angle of this rudder would increase or diminish the load on the forward or hind wheels. If we found that it would do this, but not sufficiently so, we should attach another rudder in exactly the same manner to the forward end of the machine. Suppose that, at a speed of 35 miles per hour, with both rudders set at the same angle as the aeroplane, we should find that the whole weight of the machine was carried by the aeroplane and the whole weight of the wheels (2,000 lbs.) by the track, we could then consider that the adjustment of our load was correct, and that the center of weight was directly under the center of effort for a speed of 35 miles an hour. We should then elevate the front edge of the forward rudder and depress the front edge of the rear rudder ; this would cause the machine to lift on the forward axletree and the rear end of the machine to press on the hind axletree. If we found by changing the angle of the rudders that the load could be increased or diminished on either axletree to the extent of 15 per cent. of our whole load we could consider that this phase of the problem was solved.

" For horizontal steering we should try first the effect of the screws. There should be a three-way valve in the steam pipe connected with a lever, so that we should be able to partly close off the steam from the engine of one screw, and turn more steam on to the other. This would probably be all that would be found necessary ; if not, we should try rudders.

" To prevent the machine from swaying in the air, the aeroplane should so be constructed that no matter in which direction it tilted it would diminish the lifting power of the lifted part and increase the lifting power of the depressed part. This (diedral side wings) would be simple and automatic ; moreover, the stability of the machine could be still further increased by having the center of gravity much below the center of lift.

" Having all things in readiness, the heavy wheels should

be removed and the light ones put on ; and taking one man with us to attend to the two horizontal rudders and to keep the machine on an even keel,* we should take our first fly, running the engines and doing the right and left steering ourselves. A day should be selected when there was a fresh breeze of about 10 miles per hour. We should first travel slowly around the circular railway until we came near that part of the track in which we should face the wind. The speed should then be increased until it attained a velocity of 38 or 40 miles an hour. This would lift the machine off the track and probably would slightly change the center of effort. This, however, would be quickly corrected by the man at the wheel. While the machine was still in the air careful experiments should be tried in regard to the action of the rudders ; it should be ascertained to what degree they had to be tilted in order to produce the desired effect on the machine. The machine should also be run at a speed less than 35 miles per hour in order to allow it to approach the earth gradually ; then the speed should be increased again to more than 35 miles an hour in order to rise, at the same time trying the effect of running one propeller faster than the other, to ascertain to what extent this would have to be done in order to cause the machine to turn to the right or to the left. If the machine should be constructed so that each particular foot of its surface carried a load of 1 lb. 2 oz., and if we should stop the engine dead and allow the machine to fall, it would approach the earth at a speed of 15 miles an hour, or one mile in four minutes. This evidently would cause a considerable shock, and unless there was a good deal of elasticity to the parts and a good deal of travel between the axletrees and the machine, the shock would probably be sufficient to distort or injure some part of the light structure. But it is not necessary to approach the earth directly. Professor *Langley* found in his experiments that when a horizontal plane was travelling rapidly through the air, it approached the earth as though it were ' settling through jelly.'

"A large field as near our railway as possible should be selected for alighting, and having approached the field so as to be facing the wind, we should gradually descend by slowing up the engines, and finally alight while the machine was still advancing at the rate of 20 miles an hour. If the wind should be blowing at the rate of 10 miles an hour the machine would approach the earth very gradually indeed, so that all shock would be avoided. It would only require a few yards of comparatively smooth ground

---

* M. *Maxim* has since added the gyrostat.

to run on after alighting, in order that there should be no disagreeable shock or danger.

" The cost of these experiments would be from $50,000 to $100,000, and the time required would be two years."

It will be noted how complicated and delicate these various adjustments must necessarily be, and how many different parts must be made to do their work perfectly before it can be safe to venture into the air. The aeroplane surfaces must be prevented from altering their shapes at varying speeds, the rudders must be made to maintain the course automatically, the engine must be governed as to speed, the boiler and gas-jet flames must be regulated by the consumption of steam, and the condenser must be efficient at all temperatures of the air, as well as at all speeds. Moreover, and most important, no part must break under varying strains, and the equilibrium must be maintained.

These are formidable and yet indispensable requirements, well calculated to appall the boldest inventor ; for while with an experimental model an accident is of little consequence and is easily repaired, with an actual flying machine an accident will probably prove disastrous, even if the inventor does not lose his life.

M. *Maxim*, therefore, has acted most wisely in taking plenty of time and in testing his apparatus in every way before venturing to leave the ground with it. Having completed it so that it was ready for the hazard of actual trial, he next experimented with it under conditions of comparative safety, and opened up the chapter of accidents.

The first difficulty he met with occurred through the breaking of some of the wire stays. These had been made of steel high in carbon in order to secure great tensile strength, and they proved brittle. From a private letter from M. *Maxim*, dated October 6, 1892, the writer is permitted to give the following extract, which gives also a most interesting and hitherto unpublished description of the steam-engine and boiler, which constitute thus far the great achievement of M. *Maxim :*

The steam generator is constructed somewhat on the Thorneycroft principle, except that the tubes are much lighter and thinner and have a greater number of sinuosities in them. In the Thorneycroft boiler the distributing water tubes at the bottom are of considerable size and of great weight. In my engine they are only $2\frac{1}{2}$ in. in diameter and $1\frac{1}{2}$ mm. in thickness. The downtake for the water is only 3 in. in diameter, and instead of having two, as with the Thorneycroft boiler, there is one, which branches off like the inverted letter Y. In the Thorneycroft boiler the difference in gravity of the water in

the hot interior tubes and, in the two external ones, which are
not heated, is the only means of keeping up the circulation ;
but as all the passage-ways for water are very large, this is
sufficient.

Suppose that in my system I am using steam at 300 lbs.
pressure to the square inch ; I have my water at a pressure of
335 lbs. to the square inch, and the water escapes through a
species of automatic injector, and in falling 35 lbs. in pressure
does a certain amount of work on the surrounding water.
The cold water going in from the pump is therefore made to
combine with the hot water in the downtake. This increases
the gravity of the water and at the same time causes a very
rapid forced circulation. No matter to what extent the fire
may be forced, the water has to go through in any event. All
the water that is coming in from the pump, as well as all of
the water that it takes along with it from the top separating
drum, from which the steam is taken, is forced through the hot
tubes. The nozzle through which the incoming water escapes
from the higher to the lower pressure is provided with a spring,
which always keeps a difference in pressure of about 35 lbs.;
whether the quantity of water pressing in is large or small, the
difference is always the same. A very convenient apparatus
is attached to the feed water pipe, by which it is possible to see
at a glance exactly how many pounds of water per hour are
entering the boiler. Directly over the boiler proper there is
another series of very small copper tubes through which the
water passes before entering the boiler proper, therefore products of combustion, after passing between the tubes of the
boiler, are brought in contact with the incoming water before
escaping. This so reduces the temperature of the escaping
products of combustion that Brunswick black or linseed-oil are
not burned off the smoke-stack.

For a fuel I employ naphtha of 72° Beaumé. This naphtha
is pumped into a small vertical boiler heated with a part of its
own contents.

The vapors from the boiler are led directly to an air injector,
where they escape under a pressure of 35 lbs. to the square
inch. The mixture of air and gas is then burned through rather
more than 6,000 gas jets under the boiler. Steam might be
also mixed if required. The distributing of the flame is very
even, and it is possible to fill the whole fire-box with a purple
flame. The regulating of the supply of naphtha is controlled
by the weight of the gas generator ; if the weight of the generator is too great, it operates upon a ratchet, which shortens the
stroke of the pump ; if it is too light, a spring raises the generator and its contents, when the ratchet operates in a contrary
direction and increases the stroke of the pump. In this way
the quantity of naphtha in the boiler is kept constant. The
fire is regulated not only by the pressure in the boiler, but by
a thermostatic regulator also. The feed-water pump is also
regulated by changing the length of the stroke.

The engines are compound, and have a peculiar arrangement
placed in a connection between the high and low-pressure cyl-

inders in such a manner that if the pressure in the boiler rises above 300 lbs. to the square inch the steam is shunted past the high-pressure cylinder and enters the low-pressure cylinder, and it is arranged in such a manner that the pressure of steam falling from 300 lbs. to 100 lbs. does a certain amount of work on the exhaust steam that is passing through the high-pressure cylinder after the manner of an injector—that is to say, the escaping force of the steam reduces the back pressure on the high-pressure cyl nder and increases the pressure on the low-pressure piston.

With two screws, each 17 ft. 10 in. in diameter, and with 300 lbs. pressure to the square inch, the machine has been made to pull on a dynamometer 1,960 lbs. If we multiply this pull by the number of turns per minute that the engine makes, and by the pitch of the screws, we find that the engines develop 300 horse power.

The complete weight of engines, boilers, pumps, generators, condensers, and the weight of water in the complete circulation, amounts to 8 lbs to the horse power, and this of itself I consider quite an achievement.

The spread of the wings of the machine is 107 ft., and the total length from the point of the forward rudder to the rear end of the after rudder is about 200 ft. Beneath the main aeroplane there is a considerable number of narrow planes superposed, which extend outward to nearly the full width of the machine. So far, trials have only commenced with the main aeroplane, which is 50 ft. wide and 45 ft. long in the direction of the length of the machine.

The whole machine is mounted on steel wheels 8-ft. gauge, and springs are interposed between the machine and the axletrees ; both forward and back axletrees are attached to a dynagraph, which makes a diagram of the lift of the machine as it advances upon the track. The drum which holds the paper turns once round in 1,800 ft., and whatever the machine lifts either forward or back is recorded upon the paper drum. One of the drums is also provided with a pencil, which makes a diagram of the speed at which the machine is traveling.

I am very much hampered, however, for room ; there is very little clear space between the trees, and to obtain adjoining premises without trees costs a prohibitive sum. What I should have is a circular or oval track, which would be a mile long. When the experiments are tried with a side wind blowing five miles an hour, a lift of one ton has been recorded on one side of the machine, while the other side would not lift over 100 lbs.

The whole machine, when loaded, will weigh about 7,000 lbs., so you will see if the machine will lift anything like as much, per pound of push, as I succeeded in lifting with my first apparatus, it will be sure to go.

However, I find that a great number of steel stays are necessary in order to hold the machine in shape, and while these do not weigh much, they appear to offer a considerable resistance to the passage of the machine through the air. If I were to build another machine I should aim more at getting less atmos-

pheric resistance, because I can see now that everything else is
assured except this single factor.   If the machine does not go
it will simply be because too much force is expended in driving
the framework through the air.

Work has been greatly delayed, in the first place, because
I was absent from England a great deal, and, in the second
place, we have had several serious accidents.   The high-class
steel wires—plow rope—which are used for stays are not always
reliable.   On two occasions these wires have broken, and,
becoming entangled in the wheels, have made a complete wreck
of the wheels and everything about them.   The last break-
down will take about a month to repair, and I shall put in a
lower class of steel in all the stays that are near the wheels.

This damage was duly repaired, and the experiments
were resumed early in 1893.   In one of these, with a
spread of somewhat more than half of the sustaining surface which the apparatus is designed to carry in full flight,
M. *Maxim* succeeded in obtaining, at a speed of 25 miles
per hour and with a thrust of the screws of 1,000 lbs., a
lift over the front wheels of 2,300 lbs., and over the hind
wheels of 1,900 lbs., as recorded by the dynagraphs.   On
a subsequent run, after making some alterations, he succeeded in obtaining, at a speed of 27 miles per hour and
with a thrust of screws of only 700 lbs., a lift over the
front wheels of 2,500 lbs., or quite all the weight resting
on them, and of 2,800 lbs. over the hind wheels ; thus
showing a total lift of 7 57 lbs. per pound of thrust, as
against 4.20 lbs. lifted per pound of thrust on the former
occasion.

M. *Maxim* published the diagrams illustrating both
these runs (and still another subsequently made) in the
London *Engineer* for March 17, 1893, and gave a description in which he stated that the principal lift was obtained
from the large aeroplane of 2,894 sq. ft. in area.

The run last above described was made on February 16,
1893, and on the same day two more runs were made until stopped by an accident.

First, an additional pair of wheels was attached under
the front end of the machine, connected in such a manner
that the small and lighter wheels could lift 3 in. from the
track.   Three men were also placed over the forward
axletree, and a run was then made with 900 lbs. pull on
the dynamometer.   After the machine had run about 400
ft. the light wheels lifted clear of the track, and when the
engines were stopped they came back to the track all
right.   The machine was then run again with 1,000 lbs.
pull on the dynamometer, with the following result, described in a letter to the writer from M. *Maxim*, dated
February 21, 1893 :

I have had another accident with my apparatus.

My main aeroplane is 50 ft. wide and 47 ft. long in the direction in which the machine travels. I had another aeroplane directly in front of the engine, which was about 18 ft. long and 4 ft. wide. On the first runs which I had been making I found a great deal of atmospheric resistance which I could not account for except that it resulted from the bagging of the main aeroplane and the resistance offered by the numerous struts and wires which I used in my attempts to keep it approximately flat. With the engines running at a sufficient speed to give a push of 1,325 lbs., it was found that the lift on the aeroplane did not much exceed the push of the screws.

I then made a radical change in the manner of holding the plane flat, and tried my first experiments after this with a push of 800 lbs., when it was found that the lift was a great deal more than it was with the 1.325 lbs. in the previous experiments ; in fact, the lift on the front pair of wheels was equal to the weight resting on these wheels, and the machine was only kept from leaving the track by the weight of three men whom I carried directly over the front axletree. This I regarded as dangerous. I then attached two very large cast-iron wheels in such a manner that the light wheels could lift some inches from the track before the heavy wheels were lifted at all, the weight of the heavy wheels and their axletree being about 1,400 lbs. Three men were also added to this load.

In making the run the gas was carefully turned on until the engines gave a push of 1,000 lbs. I had noticed that as the machine advanced and the engine ran faster, the boiler pressure was diminished. I therefore, upon starting, turned on a little more gas, so that the pressure, instead of falling, increased slightly during the run. When about 400 ft. had been covered, the two front wheels lifted off the track, leaving the heavy wheels still on the track ; but just before stopping the heavy iron wheels also lifted from the track, and when the engines were stopped one of the wheels got into the soft earth, sinking down and tilting the machine over to one side. A gust of wind then tipped the machine on its side ; but the breaking, which was confined almost entirely to the framework for holding the cloth, was caused by the impetuosity of a lot of men who tugged away at my ropes, and putting a strain downward instead of upward on the ropes, succeeded in completely destroying the framework.

The speed was 27 miles an hour, and the pressure of steam about 200 lbs. The lift recorded was nearly 6,000 lbs., as shown by the diagrams taken from the dynographs. The incline of the main aeroplane was, however, very steep, being about 1 in 9.

The lift was more than I expected. I did not think that a plane so very large, especially in the direction in which it was traveling, would be so efficient. I thought I should have to depend more on the narrow planes which extend beyond the main plane. This more than expected lift, however, may have been due to the wind, during the last end of the run, being contrary to the direction in which the machine was traveling.

I think that these experiments demonstrate that an aeroplane may be made to carry a considerable load.

It will take some time to repair the damage. None of the expensive machinery was damaged in the least. I shall take greater care in the future not to experiment when there is a liability to squalls, and shall have a fender, so that if the machine gets off the track it will not topple over.

It is understood that at the time this run was made about half of all the sails were in position—namely, 3,160 sq. ft. The power which the engines developed was about half of their full power, so that it will be realized that there will be ample lifting power when free flight is attempted.

Since then the apparatus has been repaired, and in an article which has been extensively published in American newspapers, a correspondent, writing under date of London, September 12, 1893, gives an account of a ride which he took on the machine. After describing it and the house in which it is sheltered, he says :

I mounted the platform, made of light matched boards so thin that they seemed scarcely able to bear a man's weight. Prior to the start a rope running to a dynamometer and post was attached behind, to measure the forward impulse or push of the screws. . . . The action of the screws caused very little shaking through the whole machine, and this was a surprise to me, comparing the tremendous force with the delicate framework. Behind the ship, 10 ft. away, two men were shouting from the dynamometer and indicating the degree of push on a large board for the engineer to read. The index quickly marked in succession 400, 500, 600, 700, and finally 1,200 lbs. of push, and then the commander yelled, " Let go !" A rope was pulled, and then the machine shot forward like a railway locomotive, and with the big wheels whirling, the steam hissing, and the waste pipes puffing and gurgling, flew over the 1,800 ft. of track. It was stopped by a couple of ropes stretched across the track working on capstans fitted with reverse fans. The stoppage was quite gentle. The ship was then pushed back over the track by the men, it not being built, any more than a bird, to fly backward.

M. *Maxim* is quoted by the correspondent as saying, among other things, concerning his apparatus :

Propulsion and lifting are solved problems ; the rest is a mere matter of time. . . . Haste in such a venture is the worst of policies. Weak points must be thoroughly sought for, and everything made completely safe before the public is invited to consider the air-ship as a practical means of transit. I am looking for a location with more room for me to experim:nt in than I can find in England. I am cramped here for want of space.

Such is the present status (1893) of this bold and costly attempt to solve the problem of aviation with an aeroplane. M. *Maxim*, as he says himself, may not achieve final success ; but he has, in the opinion of the writer, very greatly advanced the chances of eventual success. He has constructed, it may be said invented, a steam-engine with its adjuncts developing 300 horse power, and weighing only 8 lbs. to the horse power—an achievement hitherto unparalleled, and probably the most important problem to solve before man can hope to succeed in navigating the air at will.

There doubtless remain other problems to be worked out practically, notably that of effectually controlling a flying machine while in the air, both in the vertical and the horizontal direction ; that of maintaining the equilibrium under all circumstances of speed and angles of incidence, and also those of devising methods of starting up and of alighting safely anywhere ; for in practical operation, even for war purposes, M. *Maxim's* machine cannot always be brought back to get a start upon its initial railway track.

There probably also remain some questions to be settled as to the best forms, extent and texture of the supporting surfaces ; and it is not impossible that his experiments will eventually lead M. *Maxim* to a complete remodeling of his aeroplanes ; but, as has been pointed out in discussing "screws to lift and propel," it is already within his power, by reason of his marvelously light steam-engine, to go up into the air with an aerial screw, and to perform therein various evolutions.

In any event, the name of M. *Maxim* must ever remain as that of one of the men who have hitherto done most to advance the solution of the problem of aviation.

The Conference on Aerial Navigation in Chicago in August, 1893, brought out a number of experimenters whose ventures had theretofore been unpublished.

One of these, Mr. *E. C. Huffaker*, of Tennessee, had been experimenting with a model somewhat resembling the "effigy" of Mr. *Lancaster*. It consisted in a rectangular surface of fabric made concavo-convex by a rigid front spar with curved ribs at right angles thereto, so as to resemble the cross-section of a soaring bird's wing. A cross stick attached thereto carried a balancing horizontal tail, the center of gravity being determined at the front by loading with lead. The area of sustaining surface was 2 sq. ft., and when held by the cross stick at arm's length overhead, vibrating between two fingers and

facing a wind of 35 miles per hour (6 lbs. pressure at right
angles), the weight sustained (or lift) was estimated at
2 lbs. to the square foot, or that corresponding to an angle
of 10° upon a flat plane, while in point of fact the model
seemed to be horizontal, and the force required to hold it
in the wind was very small.

When the model was let go in a steady breeze it would
rise to a height of 12 or 15 ft., slowly retreating from the
wind, but always facing it ; then, tipping slightly forward,
it would descend into the face of the wind ; all these
effects being easily explained in a horizontal current.

When projected forward by hand, the model would sail
away in steady flight with a velocity of about 17 miles per
hour, and then descend on a gradient of about 1 in 15. If
thrust rapidly forward it would rise some 8 or 10 ft., and
then, hanging suspended for a moment, it sailed forward
to the ground.

These experiments are interesting as confirming what
has hitherto been said concerning the greater lift appertaining to concavo-convex surfaces, and it is to be hoped
that they will be continued.

The other experimenter was Mr. *J. J. Montgomery*, of
California. He had, some years previously, constructed
a soaring apparatus, consisting of two wings, each 10 ft.
long by an average width of 4⅓ ft., united together by a
framework to which a seat was suspended, and provided
with a horizontal tail which could be elevated or depressed
by pulleys. The wings were arched beneath, like those
of a gull, and afforded a sustaining area of about 90 sq. ft.
The weight of the apparatus was 40 lbs., and that of the
experimenter some 130 lbs. more.

Mr. *Montgomery* took this apparatus to the top of a hill
nearly a mile long, which gradually sloped at an angle of
about 10°, and placing himself within the central framework, the rods of which he grasped with each hand, ready
to sit down, he faced a sea breeze steadily blowing from
8 to 12 miles an hour, and gave a jump into the air without previous running.

He found himself at once launched upon the wind, and
glided gently forward, almost horizontally at first, and
then descended to the ground, finding that he could meanwhile direct his course by leaning to one side or the other.
The total distance glided was about 100 ft., and the sen-
sation was that of firm yet yielding and soft support, being
quite similar to the experience of M. *Mouillard*, as already
described, except that there was no apprehension of disaster.

Mr. *Montgomery* carried his machine back to the top
of the hill and prepared to repeat the experiment, but as

soon as he got into position the apparatus began to sway and to twist about in the wind ; one side dipped downward, caught on a small shrub, and, as quick as a flash, the operator was tossed some 8 or 10 ft. into the air, overturned, and thrown down headlong. He fortunately fell without serious injury, and found, as soon as he recovered himself, that one side of his machine was smashed past mending.

This experience led him to design and build a second soaring apparatus, in which he endeavored to relieve undue pressure upon either side by providing a diagonal hinge in each wing, along which the rear triangle might fold back (it was restrained by a spring) and yield to a wind gust. This apparatus measured some 132 sq. ft. of sustaining surface, and weighed 45 lbs. It was not successful ; several trials were made, but no effective lift could be obtained with it. This was attributed to the fact that the wings had been made true planes (flat) instead of being arched underneath as in the first machine.

So a third apparatus was designed and built. The wings were each 12 ft. long by an average width of 6 ft., and were given the cross-section and front sinuosity of those of a soaring vulture. They were so built and braced as to allow rotation in a socket at the front of the frame which supported the seat. A hinged tail was added, as in the two previous trials, and the machine weighed 50 lbs.

This last apparatus proved an entire failure, as no lifting effect could be obtained from the wind sufficient to carry the 180 lbs. it was designed to bear. Mr. *Montgomery* then turned his attention to other matters, but he has since made a more careful and complete study of the principles involved, and he expects to resume his experiments.

The foregoing pages comprise all the experiments, the result of which has been published, which the writer has been able to collate, and which he has considered of sufficient importance to be described in this account of " Progress in Flying Machines." Other important experiments are pending or in partial progress ; but the designers of these have as yet given out no information for publication, and indeed could scarcely do so concerning tentative plans, subject to constant modifications.

The writer has gathered from the newspapers, accounts of some other experiments, but these seem to be so erroneously or vaguely described that no instruction could be obtained by republishing them. It has been the aim of the writer throughout to gather all the information possi-

ble, but only to publish that which was reliable and instructive.

CONCLUSION.

Having thus passed in review the various attempts which
have hitherto been made to compass artificial flight, there
remains the task of pointing out as briefly as possible
whether and how the information gathered may be made
to conduce to a possible solution of the problem of aviation.

It was thought more effective to bring out the various
theories of flight, and my own views, while describing the
experiments, rather than to present them in a series of
abstract statements and propositions, the immediate bearing of which might not be so evident. The reader has
probably reached deductions of his own ; but he may also
wish to know my own general conclusions, and in what
manner if any the many failures which I have described
can be made to subserve eventual success.

These failures have resulted from so many different
causes that it is evident that many conditions must be
observed. These conditions virtually each constitute a
separate problem, which can probably be solved in more
ways than one, and these various solutions must then be
harmoniously combined in a design which shall deal with
the general problem as a whole. These various conditions, or problems, as I prefer to call them, may be enu-
merated as follows :

1. The resistance and supporting power of air.
2. The motor, its character and its energy.
3. The instrument for obtaining propulsion.
4. The form and kind of the apparatus.
5. The extent of the sustaining surfaces.
6. The material and texture of the apparatus.
7. The maintenance of the equilibrium.
8. The guidance in any desired direction.
9. The starting up under all conditions.
10 The alighting safely anywhere.

Analyzed and viewed in this way, the reader may realize
how complicated is the question and how formidable are
the various difficulties which are to be surmounted. And
yet the scrutiny which has been made of the various experiments attempted and of the progress accomplished in
flying machines enables us to perceive that many of these
problems have been approximately solved, more particularly since 1889, and that a better understanding of the
difficulties to be overcome has been obtained concerning
several others.

1. The first problem to be considered is that pertaining

to the resistance and supporting power of air. By the use of currently accepted formulæ it could not be figured out a few years ago how birds were supported in flight. Now that Professor *Langley's* experiments have confirmed many of those previously tried, we are enabled to say that the empirical formula of *Duchemin* (from which the table of " lift" and " drift" herein given was calculated) is approximately correct, and to figure out the support and the resistance with some confidence of not going far wrong.

These calculations seem to indicate that artificial flight is possible, even with planes ; that very flat angles of incidence, from 2° to 5°, hitherto considered inadmissible, will be the most advantageous, and that within certain limits of hull resistance high speeds will require less power than low speeds, because they admit of obtaining support from the air at a flatter angle.

We have seen that the " drift" diminishes as the angle of incidence becomes less, that the " hull resistance" (including car, framing, braces, etc.) increases as the square of the speed, and that the skin friction is so small that it may for the present be disregarded, and we are enabled to calculate, approximately at least, the power required to obtain support in flight with planes, and to overcome the resistance, although we are not yet aware what limit will be imposed upon the size of artificial apparatus by the law that the weight will increase as the cube, while the sustaining surfaces will grow only as the square of the similar dimensions.

Moreover, the formulæ which give this promise of success were derived from experiments with plane surfaces, and we already know that concavo-convex surfaces will be still more effective, although the most favorable shapes are not yet ascertained. This statement indicates the direction in which scientific investigation and experiment should now proceed, and holds out the hope that this first problem is in a fair way of being solved.

2. The second problem—that concerning the motor to be employed—has justly been considered to be the most important and difficult of solution. It seemed hopeless to rival, with an artificial motor, the output of energy appertaining to the motor muscles of birds in proportion to their weight, which, as we have seen, there is good reason to believe develop work in ordinary flight at the rate of 1 H.P. to 20 lbs. of weight, and can for a brief period, in rising, give out energy at such rate as to represent an engine of only 5 or 6 lbs. of weight developing 1 H.P.

The writer has, on a former occasion,* passed in review

---

* Aerial Navigation. A lecture to the students of Sibley College, 1890.

the comparative weights of various classes of engines. He found that the lightest engines in use in 1890, including the generator of power, weighed 60 lbs. per H.P. for steam, 88 lbs. per H.P. for gas engines, and 130 lbs. per H.P. for electric motors. He intended to discuss the subject further in this account of " Progress in Flying Machines," but recent achievements with steam-engines seem to make this unnecessary. Marine (yacht) engines have been reduced more than one half in weight ; Mr. *Hargrave* has produced a steam-engine weighing 10.7 lbs. per H.P.; M. *Maxim* has created one weighing but 8 lbs. per H.P., including a condenser, and other experimenters are approximating closely to the same weights.

Steam-engines, therefore, seem to have been so much reduced in weight as to admit of their being employed as motors for flying machines. This may not be a final solution, for it may be that some form of gas or petroleum engine will prove to be still better adapted to aerial purposes, as indeed has been already hinted by M. *Maxim ;* but in any event, his steam-engine seems to be light enough to make a beginning of artificial flight, if the other problems pertaining thereto can also be solved.

But it is possible to utilize a still lighter power, for we have seen that the wind may be availed of under favorable circumstances, and that it will furnish an extraneous motor which costs nothing and imposes no weight upon the apparatus.

Just how much power can be thus utilized cannot well be told in advance of experiment ; but we have calculated that under certain supposed conditions it may be as much as some 6 H.P. for an aeroplane with 1,000 sq. ft. of sustaining surface ; and we have also seen that while but few experimenters have resorted to the wind as a motor, those few have accomplished remarkable results.

3. As regards the selection of the instrument through which propulsion is to be obtained, we have seen that experiment has shown that reaction jets, whether obtained from explosives, steam, or blasts of air ; that wave action ; that valvular, folding or feathering paddles or vanes have all proved inferior in practical application to screw propellers or to propelling wings, and that the two latter (if we are to judge from Mr. *Hargrave's* experiments) are about equally effective. It being understood, however, that this statement refers to wings only as propelling instruments and not as sustaining surfaces. We may conclude, therefore, that the third problem may now be solved either with screws or with waving wings, as best conforms to the rest of the design.

4. This brings us, therefore, to consider the solution of

the fourth and important problem of what kind or form of apparatus should be selected for sustaining the weight—whether flapping wings, screws, or aeroplanes. The best measure of comparison will be the weights or number of pounds which experiment shows may be sustained per H.P. with each form, considered in connection with the weight of the construction required to make that form abundantly strong against the resulting strains. The difference between the two weights will indicate the proportion of the whole which may be devoted to the motor. It is desirable, therefore, to consider each form or kind of apparatus separately.

We do not yet know accurately how many pounds per H.P. can be sustained in horizontal flight with a bird-like apparatus of flapping wings. The toy birds which have been described support only from 6 to 20 lbs. per H.P., but this inefficiency is largely due to the undue friction of the working parts and to the abnormal head resistance of the framing in such imperfect models. The writer has estimated that in the case of a flying pigeon about 77 lbs. are sustained per H.P., but as this is partly based on conjecture, it may be an underestimate.*

Upon the whole, the writer is inclined to admit that about 100 lbs. per H.P. may be sustained with flapping wings, this including the power required both to support the weight and to overcome the head resistance. He believes, moreover, that in an artificial machine of sufficient size to sustain one man, the strength required to resist the constant reversals of strains due to the alternating motion of the wings will involve such dimensions that the weight of the apparatus and man will amount to at least three-fourths of the whole, thus leaving but one-fourth of the total weight which can be devoted to the motor and its adjuncts, including the fuel and supplies for the journey.

Concerning aerial screws we have abundant experimental data. *Nadar, Wenham* and *Freninges* each obtained a sustaining effect of 33 lbs. per H.P.; *Dieuaide* realized 26.4 lbs., and *Dahlstrom* and *Lohman* secured 37.6 and 55 lbs. per H.P., while *Renard* obtained from 17 to 48 lbs. thrust by screws rotating at various speeds, and *Moy* recorded 40 lbs. per H.P. sustained from a wind wheel with vanes of variable pitch.

These performances, however, included a certain amount of ascension, which absorbed part of the power, so that probably we shall be quite safe in assuming that

---

* M. *Lilienthal* raised 93 lbs. per H.P. in his experiments, but this was counterbalanced weight, and there was no head resistance of forward flight to be overcome.

in mere horizontal flight some 45 lbs. per H.P. can be sustained with screws.

As the strains in a rotating apparatus will be less destructive than those involving reversals of motion, it seems probable that screws may be constructed with a less weight of materials than flapping wings of the same sustaining power. It is judged that an apparatus can be constructed to sustain the weight of one man with rotating screws, in which only about two-thirds of the weight shall be absorbed by the framing, screws, car and man, thus leaving one-third of the whole weight for the motor and its various adjuncts. The practical result of this estimate will be elicited further on.

We have also a number of experimental data concerning aeroplanes. Professor *Langley* sustained a maximum of 209 lbs. per H.P. with planes at an angle of incidence of 2°, and M. *Maxim* sustained 133 lbs. per H.P. at an inclination of 1 in 14. These data apply to the plane only. Neither of these performances included the head resistance due to the framing and car which are indispensable in an actual machine, so that we must derive our premises from complete models. With one of the latter *Tatin* sustained 110 lbs.; *Phillips*, 72 lbs., and *Hargrave*, 89 lbs., 76 lbs., and 79 lbs. per H.P. in horizontal flight. We may safely conclude, therefore, that 100 lbs. per H.P. can be sustained in horizontal flight with an aeroplane.

As the latter consists of fixed surfaces receiving no strains save the sustaining pressure of the air, it is believed that such class of apparatus can be constructed of sufficient size to sustain one man, so that about one-half of the whole weight shall be devoted to the apparatus and man, and the other half to the motor and its adjuncts.

These estimates of the proportion of the sustained total weight which can be spared for the motor, are necessarily mere estimates made in advance of actual testing, and (for reasons to be stated hereafter) upon the smallest size of apparatus practicable for actual man flight, yet they enable a comparison to be made between the various forms of apparatus which have been herein described. The result is as follows :

COMPARATIVE EFFICIENCY OF VARIOUS FORMS.

Kind of Apparatus.	Pounds sustained per H.P.	Proportion available for motor.	Resulting possible weight of motor per.H.P.
Screws....................	45	$\frac{1}{3}$	15 lbs.
Wings..	100	$\frac{1}{4}$	25 "
Aeroplanes.............	100	$\frac{1}{2}$	50 "

The above table, based as it is upon experimental data of weights actually sustained, indicates that aeroplanes are probably the best forms to experiment with, because they admit of a larger proportion of the whole weight being appropriated to the motor. It also indicates the possibility of success in artificial flight, with motors weighing 10 or 15 lbs. per H.P., provided that the remaining problems be also solved ; but it must not be overlooked that more power will be required in rising from the ground than in horizontal flight, and that the actual proportion of the total weight available for the motor, although conservatively estimated from the best data available,* is still a matter to be proved by experiment.

The common basis which has been here selected for comparison is that size of apparatus sufficient to support the weight of one man. This is the smallest which can be adopted, and it is theoretically the most favorable, for inasmuch as the weight of the framing will presumably increase as the cube of the dimensions, while the sustaining surfaces will increase as the square of these same dimensions, it is seen that the ratio of the total weight sustained which can be spared for the motor will not be constant, but that the larger the apparatus the more it will weigh in proportion to its surface and the less there will remain for the engine and its adjuncts. Flying machines, therefore, should preferably be designed as small as practicable, and experimenters will place themselves at a disadvantage if they construct large machines.

5. As regards the fifth problem—the amount of the sustaining surface required—it depends on the speed, and it is probable that, within certain limits, no particular extent (in ratio to the weight) can be said to be absolutely the best, because a large part of the resistance will consist in the " drift," and the latter is independent of the area of the sustaining surface ; a small area at high speed being able to sustain as much weight as a larger area at a corresponding lesser speed, as indeed is indicated by the formula already given for the drift : $R = W \tan g. @$, in which the element of surface disappears.

Practically, however, the weight of the necessary framing and the hull resistance will determine the ratio of surface to weight which will be most advantageous. We have seen that encouraging experiments have been made with surfaces varying from 0.75 sq. ft. per pound in the case of Herr *Lilienthal* to 7 sq. ft. per pound in the case of M. *Hargrave.* It seems probable that the latter is in excess, and that it would be preferable to confine the

---

* *Maxim's* aeroplane and the soaring devices of *Le Bris, Mouillard, Lilienthal*, and *Montgomery.*

dimensions of artificial machines within the proportions which obtain with fast flying birds, as shown in the table heretofore given, this being from 3.62 sq. ft. per pound in the case of the swallow, to 0.44 sq. ft. per pound in the case of the male duck, with which areas, if we consider their wings as planes, and the angle of incidence to be $3°$, the swallow requires a speed of 23.1 miles per hour and the duck a velocity of 66.2 miles per hour to sustain their weight.

To come down safely, at the speed of the parachute, requires about the ratio of the swallow, while the proportions of the duck are more favorable to high speed. As the drift will increase only if the angle of incidence be increased, it would seem preferable to maintain this angle as uniform as possible, and to provide variable supporting surfaces to be folded or unfolded with variations of speed, if such a construction can be devised in connection with the concavo-convex surfaces which have already been mentioned as likely to give the most satisfactory results.

6. The sixth problem cannot be said to be solved, for there is considerable uncertainty concerning the best materials to be employed for the framing and for the moving parts ; or what should be the texture of the sustaining surfaces in an actual flying machine. Hitherto the main question has been to construct a model which would fly at all ; and experiments with models have not thrown much light on the question of materials. If a partial success be realized, this problem will assume greater importance.

It involves considering materials from a somewhat new point of view, or investigating their strength and stiffness per unit of weight, so as to secure a maximum of resistance with a minimum of weight. The quill of a bird's feather is stronger and more elastic than an equal weight of steel, and the texture of its barb is peculiar.

It now seems probable that bamboo, the lighter of the stiff woods, and some varieties of steel, will be found to be the preferable materials for the framing. Contrary to popular belief, aluminium is inferior to steel per unit of weight, particularly in compression, but it does not corrode and may be preferable on that account. It may be utilized for the sustaining surfaces, either as thin sheets or as wire gauze made smooth by some coating ; but textile fabrics will probably be the first to be employed for full-sized apparatus. One important requirement, however, is that the surfaces shall not unduly change their shape under varying air pressures. They must be rigid, and, perhaps, elastic, and the fluttering of textile fabrics is likely to give trouble to experimenters. It may be, therefore, that thin wood, parchment, or pasteboard may prove preferable, the latter being corrugated lengthwise of the direction of motion in order to gain stiffness.

The barb of a feather is smooth in one direction and asperous in the other ; and it is possible that a similar texture of surface may prove of advantage in flying machines, but this probably will not be determined until partial success has been achieved with an apparatus of sufficient size to sustain the weight of a man.

7. The problem of the maintenance of the equilibrium is now, in my judgment, the most important and difficult of those remaining to be solved. It has been seen, from this review of "Progress in Flying Machines," that almost every failure in practical experiments has resulted from lack of equilibrium. This is the first requisite thing to secure, for, as has already been said, safety is the most important element of success—safety in starting up, in sailing, and in coming down.

If a flying machine were only required to sail at one unvarying angle of incidence in calm air, the problem would be much easier of solution. The center of gravity would be so adjusted as to coincide with the center of pressure at the particular angle of flight desired, and the speed would be kept as regular as possible ; but the flying machine, like the bird, must rise and must fall, and it must encounter whirls, eddies, and gusts from the wind. The bird meets these by constantly changing his center of gravity ; he is an acrobat, and balances himself by instinct ; but the problem is very much more difficult for an inanimate machine, and it requires an equipoise—automatic if possible—which shall be more stable than that of the bird.

We have seen from the experiments described that the transverse stability can be procured in two ways : (1) by placing the two halves of the sustaining surfaces at a diedral angle to each other, and (2) by adding a longitudinal keel to the apparatus, as in the case of Mr. *Boynton's* fin kites. The mode of action is practically the same for both, and consists in producing increased air pressure upon the side which tends to dip downward. The two may be employed conjointly, but the keel will produce less head resistance to forward motion than the diedral angle, which resistance, however, may be diminished by turning upward only the outer ends of the sustaining surfaces in a manner similar to the upbending primary feathers of the soaring birds.

Longitudinal stability may be promoted in three ways : (1) By additional surfaces at a slight angle to the main sustaining surface, (2) by placing several surfaces behind each other, (3) by causing the center of gravity always to coincide with the center of pressure. The first way corresponds to the method which has been mentioned as procuring transverse stability by means of surfaces at a diedral angle ; it is illustrated by M. *Maxim's* aeroplane, in which two such surfaces are affixed, front and rear ; and by M. *Pénaud's* aeroplane, in which but one is affixed in the rear. The second way is illustrated by M. *Brown's* bi-planes and by M. *Hargrave's* cellular kites ; and the third is the method universally employed by the birds.

For an artificial machine this last method is as yet an unsolved problem. Several inventors have proposed methods of shifting weights to change the position of the center of gravity as the apparatus changes its angle of incidence, but none of these are automatic, and none have been tested practically.

8. The guidance in a vertical direction—*i e.*, up or down, depends in a great degree upon success in the changing the center of gravity which has just been alluded to. It may be partly effected by changes in the speed or by horizontal rudders, but in such case the equilibrium will be disturbed. Guidance in a horizontal direction has been secured, as we have seen in several experiments, by vertical rudders ; but there are probably other methods still more effective, although their merits cannot be tested until a practical apparatus is experimented with. Upon the whole, this problem may give trouble, but it does not seem unsolvable.

9. A really adequate practical flying machine will hardly be said to have come into existence until it possesses the power of starting up into the air under all conditions. This problem is as yet unsolved, and may not be until the other problems have been worked out to a success. It is clear that in rising upward more power will be required than in horizontal flight ; for to the force required to obtain horizontal support must be added that required to ascend, and the latter will vary with the rapidity of the upward motion. Three principal methods have been experimented with : (1) By acquiring speed and momentum on the ground ; (2) by the reaction of rotating screws ; (3) by utilizing the force of the wind. The first we have seen to require the use of special appliances, such as railway tracks, so that its application must be limited, and the third necessitates that the wind shall blow, and with sufficient force ; either or both may be utilized with the earlier types of practical machines should one or more be hereafter developed ; but the writer believes that the second method—that of rising through the reaction of a screw —will eventually supersede the two others. It will in-

volve the difficult design of a simple form of sustaining surfaces which can be alternately rotated as a screw or held as a fixed aeroplane when sailing, the change being effected while under motion in the air.

The writer does not believe that a bird-like machine can rise into the air, under all conditions, by flapping its artificial wings. It would need to be already up some distance to permit such action. Birds spring up three or four times their own height, or run against the wind to acquire speed, and with vigorous flaps of wing they rise at an angle seldom greater than 45°, but their initial action would be quite impracticable to a machine of sufficient size to sustain the weight of a man.

10. The alighting safely anywhere is also an unsolved problem, and one, as will readily be perceived without argument, of vital consequence. It has been slurred over by most of the designers of flying machines, and the best method which has been thus far proposed involves the selection of a smooth, soft piece of ground and the alighting thereon at an acute angle. When it is considered that the speed required for support will be somewhere from 20 to 40 miles an hour, it will be realized that the performance will be somewhat dangerous, and that it would be preferable, if the design of the apparatus will admit of it, to imitate the manœuvre of the bird who stops his headway by opening his wings wide, tilting back his body, and obtaining the utmost possible pressure and retardation from the air before alighting upon the ground. This would require for an artificial machine a rapid change of the center of gravity so as to tilt the apparatus backward to the angle of maximum lift (about 36° by the table) and, immediately thereafter, a counter change of the center of gravity, so as to bring the apparatus back upon an even keel in order to alight at the diminished velocity.

This manœuvre is not as difficult and dangerous as may at first sight appear ; but it must be acknowledged that it would be preferable to utilize the reaction of a rotating screw to diminish the forward motion and to hover over the ground before alighting. This involves the same difficult design which has been alluded to as desirable for use in rising, for it does not seem practicable, within the requisite limits of weight, to provide two sets of sustaining surfaces, one set to be used in rising and in alighting, and the other to serve in horizontal flight. These last two problems—the rising and the alighting safely, without special preparation of the ground—seem very difficult of solution, and are probably the last which will be worked out.

The general problem having been thus decomposed into

its several elements, and each element considered as a separate problem, it will be seen that the mechanical difficulties are very great ; but it will be discerned also that none of them can now be said to be insuperable, and that material progress has recently been achieved toward their solution.

The resistance and supporting power of air are approximately known, the motor and the propelling instrument are probably sufficiently worked out to make a beginning ; we know in a general way the kind of apparatus to adopt, its approximate extent and required texture of sustaining surfaces, and there remain to solve the problems of the maintenance of the equilibrium, the guidance, the starting up, and the alighting, as well as the final combination of these several solutions into one homogeneous design.

In spite, therefore, of the continued failures herein recorded, it is my own judgment, as the result of this investigation into the " progress in flying machines," more particularly the progress of late years, and into the recent studies of the principles and problems which are involved, that, once the problem of equilibrium is solved, man may hope to navigate the air, and that this will probably be accomplished (perhaps at no very distant day), with some form of aeroplane provided with fixed concavo-convex surfaces, which will at first utilize the wind as a motive power, and eventually be provided with an artificial motor.

The conclusion that important progress may be achieved without an artificial motor was little expected when this investigation was begun ; but the study of the various experiments which have been passed in review, the perception of the partial successes which have been accomplished with soaring devices, and the general consideration of the subject, have led to the conclusion that the first problem which it is needful to solve is that of the equilibrium, and that in working this out the wind may furnish an adequate motive power.

Preliminary experiments will, of course, be tried upon a small scale, but no experiment with a model can be deemed quite conclusive until the same principles have been extended to a full-sized apparatus capable of sustaining a man, and until this has been exposed to all the vicissitudes of actual flight. It will readily be discerned that a less achievement than this would not prove an adequate performance, and that no matter how well a model might behave in still air, there would still remain the questions as to how it would behave in a wind, and how it was to solve the problems of starting up and of alighting.

It would seem, therefore, that the first problem to solve is that of the maintenance of the equilibrium at all the

angles of incidence required ; in rising, in sailing, in en-
countering wind eddies, and in alighting.

For this purpose, it is now my opinion, based upon the
performance of the soaring birds and upon the partial success of some soaring devices, that the problem of equilibrium can best be solved with an apparatus which shall
utilize the wind as a motive power—*i.e.*, with some form
of aeroplane of sufficient size to sustain a man, with which
the operator shall endeavor to perform the various manœuvres required to meet the varying conditions of actual
flight, and to preserve at all times his balance in the air.
In other words, a flying machine to be successful must be
at all times under intelligent control, and the skill to obtain that control may be acquired by utilizing the impulse
of the wind, thus eliminating, for a time at least, the
further complications incident to a motor.

But whether a soaring device be first experimented with,
or whether the initial apparatus be provided with a motor,
the next question pertains to the conditions under which
a machine carrying a man can be experimented with most
safely.

Various methods have been suggested, and a few have
been tried. The most obvious is to suspend the apparatus
from a cable stretched between two tall masts or between
two steep hills. This has been proposed many times, but
we have seen by the experience of M. *Sanderval* that it
does not afford sufficient length of suspending rope to permit of unimpeded manœuvres, and that experience gained
in that way would scarcely be available in free flight. A
preferable plan has been proposed by M. *Duryea* (and
probably by others), which consists in suspending the apparatus to be experimented with from a captive balloon,
anchored by several divergent ropes so as to remain a half
mile or thereabouts from the ground, as shown in fig. 83.
By means of a rope passing through a pulley block attached to the balloon, and thence to a windlass on the
ground, the machine to be experimented with may be
drawn into the air to a sufficient height to clear the gusty
air conditions found at or near the ground, and there, in
comparative safety, the sky-cycler might manipulate his
devices, ascertain the effect of various manœuvres, and
gradually gain control, skill, and confidence preparatory
to trusting himself to actual flight.

This method is understood to have been employed by
M. *C. E. Myers*, the aeronautical engineer, in experimenting with parachutes, and to have given promise of
satisfactory results within certain limits. It is well worth
testing as a preliminary trial of a flying apparatus, but it
should be remembered that a machine suspended from a

rope, however long, will not be under quite the same circumstances as in free flight. Even if it rises upon the
wind and is wholly supported thereon it will still be hampered by the rope, and perhaps restrained from some
action which it is important to understand in order to
maintain the equilibrium, so that the operator will never
be quite certain that he has gained complete control over
his apparatus.

Other methods have been proposed by various writers.

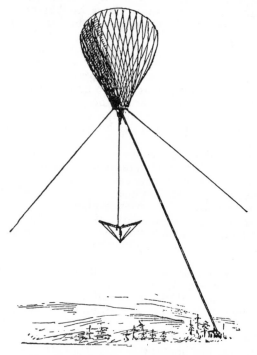

FIG. 83.—DURYEA'S PROPOSAL—1893.

M. *Ch. Weyher*, for instance, in the *Aéronaute* for July,
1884, suggested the construction of a circular railway of
600 to 1,000 ft. diameter, upon which a large platform car,
covered with a soft mattress, should carry the apparatus
to be tested, attached with restraining ropes about breast
high. This car to be towed at varying speeds by a locomotive, so as to afford a sustaining effect and to encounter
the wind at various angles, until the operator shall master
the necessary manœuvres.

M. *A. Goupil*, on the other hand, proposes a circular

elevated railway consisting of a single central girder suspended by wire ropes between two rows of posts, and serving to carry a truck to which the apparatus to be experimented with may be suspended. In this case the machine might be provided with its own motive power, or towed by a wire rope, or driven by an electric motor, but in either case there would still remain the restraint of the safety suspending rope, which, as previously suggested, might vitiate the various air reactions which it is important for the operator to experience practically.

These, and other devices which may be suggested, may doubtless prove useful in making the preliminary trials with various forms of apparatus, thus testing their behavior when restrained, but there will always come a time when such apparatus, if apparently adequate, must be tried at full liberty and encounter all the contingencies of free flight. It seems clear that after the preliminary trials with models have been made, time may be saved in ascertaining the full merits of a device and in improving it, if experiments with the full-size apparatus be made at entire liberty instead of under restraint, provided adequate precautions be taken to avoid serious injury in alighting.

Referring to the various experiments which have been made with full-sized apparatus, more particularly those of *Dante, Le Bris, Mouillard, Lilienthal,* and *Montgomery,* it is seen that *Dante* adopted the more rational plan of all by experimenting over a sheet of water, although the exact method he pursued is not known.

Upon the whole, the best mode of procedure is probably that proposed by *Le Bris,* which want of means prevented him from adopting—that is to say, to start from the deck of a steam vessel under way, so as to obtain initial velocity, as well as to face the wind from whatever direction it may blow, and to be quickly picked up after alighting. If the machine be provided with a light buoy and line, and the operator be encased in a cork jacket or life-preserver, he may thus quickly put to the test the merits and the deficiencies of his apparatus with but little danger to himself, and ascertain whether it can be brought under control. The machine may experience breakages, the operator will doubtless suffer many duckings, he may even be stunned at times, but he is not likely to lose his life or to break a limb, as he might do were he to experiment over land.

It is believed that salt water is preferable to fresh water, over which to carry on such experiments, not only because of the greater buoyant power of the water, but especially because sea breezes are more regular and less gusty than land breezes. It is evident that it would be preferable to operate over a genial or a tepid sea, in trade-wind regions if possible, and in locations where steady sea breezes of no great intensity may be relied upon to blow almost daily. It would be desirable to select the vicinity of some projecting tongue of land or of some isthmus, where captive preliminary tests may be made, and also that there should be a cliff in the neighborhood whence models and perhaps the apparatus itself might be floated off. There are many such spots to be found within proximity of machine shops, in the Mediterranean, in the Gulf of Mexico, and on the coast of Southern California, and the attention of designers of flying machines, who may want to test the merits of their devices upon a really adequate scale, is particularly directed to the vicinity of San Diego, Cal., where all the circumstances which have been alluded to are to be found combined, even to a local railroad along the beach, on which the tests proposed by M. *Weyher* might be carried on.

All this presupposes that the preliminary experiments with small models have resulted satisfactorily, and that the designer wishes further to test the merits of his apparatus upon a practical working scale, with a machine capable of carrying a man and provided with the requisite devices to bring it under control while in the air, and thus to work out the problem of equilibrium. The expense will doubtless be considerable, and the mishaps not infrequent, but there seems to be no surer way of ascertaining whether a full-sized apparatus will preserve its balance in the air, while the risk of serious injury will be small. If such experiments finally succeed in solving the equilibrium problem, in securing safety in rising, in sailing, and in coming down, with a machine carrying its operator, an immense step forward will have been taken toward solving the other problems mentioned, and toward finally developing a safe flying machine, provided with a motor of its own and capable of being operated anywhere ; for once safety has been secured under the various actual conditions of out-door performance, it ought to be a comparatively easy and short task to work out the other questions, save perhaps those pertaining to the starting up from and alighting upon the ground.

Assuming all this to be possible—and while the mechanical difficulties are doubtless great, they do not seem to be insuperable—the final working out of the general problem is likely to take place through a process of evolution. The first apparatus to achieve a notable success will necessarily be somewhat crude and imperfect. It will probably need to be modified, reconstructed, and readventured many times before it is developed into practical shape.

The inventor will doubtless have to construct the first models and perhaps the first full-sized machine at his own expense, in order to demonstrate the soundness of his conception and the comparative safety of its operation ; but after this much is accomplished further remodelling and experiment will still probably be required to develop the apparatus into commercial value.

This phase of the evolution is likely to require the aid of capital, because the expense may be quite considerable ; and inasmuch as a financial venture to develop such a difficult and novel contrivance must be gone into as a hazard, with the acceptance of the possibility of total loss as an offset for the hope of drawing a prize, the parties advancing the capital will probably require that the invention (if invention there be) shall be fully protected by patents.

In view of this probable requirement, it may be questioned whether M. *Hargrave* is quite prudent in taking out no patents for his various devices, for he hints in his last paper that he is hampered in his experiments by having to perform them in public. The difficulty arises from the fact that the experiments, to be of practical value, have to be performed out of doors, and the writer knows of some designers who, unable, on the one hand, to secure a patent—in the United States at least—until they can demonstrate the practical performance which they hope for, and apprehensive, on the other hand, of being annoyed by spectators, have retired into a wilderness to make their experiments, thus placing themselves at serious disadvantage in case a mishap of any kind occurs.

Most of the patents heretofore granted for flying machines are quite impracticable, yet the claims cover, here and there, some feature which may eventually contribute to success. It will be judicious, therefore, for designers of projected flying machines to study prior patents, and an attempt has been made in these pages to indicate some of those which contain valuable suggestions. The novelty (if any) in future patents will probably largely consist in new combinations of features already patented.

There are probably a good many arrangements of sustaining surfaces which will prove available for aeroplanes ; some will prove more effective and steadier than others, and this must be ascertained by experiment ; but in any event success would be hastened by a working association of experimenters in this inchoate research, for the problems, as has been seen, are many, and no inventor is likely to be in possession of all the miscellaneous knowledge and variety of talent required to perfect so novel an undertaking.

To the possible inquiry as to the probable character of a successful flying machine, the writer would answer that in his judgment two types of such machines may eventually be evolved : one, which may be termed the soaring type, and which will carry but a single operator, and another, likely to be developed somewhat later, which may be termed the journeying type, to carry several passengers, and to be provided with a motor.

The soaring type may or may not be provided with a motor of its own. If it has one this must be a very simple machine, probably capable of exerting power for a short time only, in order to meet emergencies, particularly in starting up and in alighting. For most of the time this type will have to rely upon the power of the wind, just as the soaring birds do, and whoever has observed such birds will appreciate how continuously they can remain in the air with no visible exertion. The utility of artificial machines availing of the same mechanical principles as the soaring birds will principally be confined to those regions in which the wind blows with such regularity, such force, and such frequency as to allow of almost daily use. These are the sub-tropical and the trade-wind regions, and the best conditions are generally found in the vicinity of mountains or of the sea.

This is the type of machine which experimenters with soaring devices heretofore mentioned have been endeavoring to work out. If unprovided with a motor, an apparatus for one man need not weigh more than 40 or 50 lbs., nor cost more than twice as much as a first-class bicycle. Such machines therefore are likely to serve for sport and for reaching otherwise inaccessible places, rather than as a means of regular travel, although it is not impossible that in trade-wind latitudes extended journeys and explorations may be accomplished with them ; but if we are to judge by the performance of the soaring birds, the average speeds are not likely to be more than 20 to 30 miles per hour.

The other, or journeying type of flying machines, must invariably be provided with a powerful and light motor, but they will also utilize the wind at times. They will probably be as small as the character of the intended journey will admit of, for inasmuch as the weights will increase as the cube of the dimensions, while the sustaining power only grows as the square of those dimensions, the larger the machine the greater the difficulties of light construction and of safe operation. It seems probable, therefore, that such machines will seldom be built to carry more than from three to 10 passengers, and will never compete for heavy freights, for the useful weights, those carried in addition to the weight of the machine itself, will be very small in proportion to the power required. Thus M. *Maxim* provides his colossal aeroplane (5,500 sq. ft. of surface) with 300 horse power, and he hopes that it will sustain an aggregate of 7 tons, about one-half of which consists in its own dead weight, while the same horse power, applied to existing modes of transportation, would easily impel—at lesser speed, it is true—from 350 to 700 tons of weight either by rail or by water.

Although it by no means follows that the aggregate cost of transportation through the air will be in proportion to the power required, the latter being but a portion of the expense, it does not now seem probable that flying machines will ever compete economically with existing modes of transportation. It is premature, in advance of any positive success, to speculate upon the possible commercial uses and value of such a novel mode of transit, but we can already discern that its utility will spring from its possible high speeds, and from its giving access to otherwise unreachable points.

It seems to the writer quite certain that flying machines can never carry even light and valuable freights at anything like the present rates of water or land transportation, so that those who may apprehend that such machines will, when successful, abolish frontiers and tariffs are probably mistaken. Neither are passengers likely to be carried with the cheapness and regularity of railways, for although the wind may be utilized at times and thus reduce the cost, it will introduce uncertainty in the time required for a journey. If the wind be favorable, a trip may be made very quickly ; but if it be adverse, the journey may be slow or even impracticable.

The actual speeds through the air will probably be great. It seems not unreasonable to expect that they will be 40 to 60 miles per hour soon after success is accomplished with machines provided with motors, and eventually perhaps from 100 to 150 miles per hour. Almost every element of the problem seems to favor high speeds, and, as repeatedly pointed out, high speeds will be (within certain limits) more economical than moderate speeds. This will eventually afford an extended range of journey —not at first probably, because of the limited amount of specially prepared fuel which can be carried, but later on if the weight of motors is still further reduced. Of course in civilized regions the supply of fuel can easily be replenished, but in crossing seas or in explorations there will be no such resource.

It seems difficult, therefore, to forecast in advance the commercial results of a successful evolution of a flying

machine. Nor is this necessary; for we may be sure
that such an untrammelled mode of transit will develop a
usefulness of its own, differing from and supplementing
the existing modes of transportation. It certainly must
advance civilization in many ways, through the resulting
access to all portions of the earth, and through the rapid
communications which it will afford.

It has been suggested that the first practical application
of a successful flying machine would be to the art of war,
and this is possibly true; but the results may be far different from those which are generally conjectured. In the
opinion of the writer such machines are not likely to prove
efficient in attacks upon hostile ships and fortifications.
They cannot be relied upon to drop explosives with any
accuracy, because the speed will be too great for effective
aim when the exact distance and height from the object
to be hit cannot be accurately known. Any one who may
have attempted to shoot at a mark from a rapidly moving
railway train will probably appreciate how uncertain the
shot must be.

For reconnoitring the enemy's positions and for quickly
conveying information such machines will undoubtedly
be of great use, but they will be very vulnerable when
attacked with similar machines, and when injured they
may quickly crash down to disaster. There is little question, however, that they may add greatly to the horrors
of battle by the promiscuous dropping of explosives from
overhead, although their limited capacity to carry weight
will not enable them to take up a large quantity, nor to
employ any heavy guns with which to secure better aim.

Upon the whole, the writer is glad to believe that when
man succeeds in flying through the air the ultimate effect
will be to diminish greatly the frequency of wars and to
substitute some more rational methods of settling international misunderstandings. This may come to pass not
only because of the additional horrors which will result
in battle, but because no part of the field will be safe, no
matter how distant from the actual scene of conflict. The
effect must be to produce great uncertainty as to the results of manœuvres or of superior forces, by the removal
of that comparative immunity from danger which is necessary to enable the commanding officers to carry out their
plans, for a chance explosive dropped from a flying machine may destroy the chiefs, disorganize the plans, and
bring confusion to the stronger or more skillfully led side.
This uncertainty as to results must render nations and
authorities still more unwilling to enter into contests than
they are now, and perhaps in time make wars of extremely
rare occurrence.

So may it be ; let us hope that the advent of a successful flying machine, now only dimly foreseen and nevertheless thought to be possible, will bring nothing but good into the world ; that it shall abridge distance, make all parts of the globe accessible, bring men into closer relation with each other, advance civilization, and hasten the promised era in which there shall be nothing but peace and good-will among all men.

THE END.

APPENDIX.

THE FLIGHT OF THE ALBATROSS.

PAPER READ BEFORE THE BALLOON SOCIETY, OCTOBER 3, 1884.

By Thomas Moy.

BEFORE I describe the mechanical portion of my subject it may be useful if I make a few remarks on matters which, although personal to myself, have now become historical.

In October, 1859, I published in the *Mechanic's Magazine* a description of a cigar-shaped vessel to contain gas for support, and propellers placed in the same plane as the resistance, not yards below, as our French friends persist in placing the power. I subsequently concluded that the bird principle proffered greater chances of success.

Then on May 20, 1869, I described the flight of the albatross, which I now intend to treat upon. Bear in mind that this subject is more than 15 years old. The only result arising therefrom was a word or two at the meeting, in opposition, in the discussion that followed.

Well, in 1875 I built a flying machine for experiment. I made a steam-engine which gave an actual 3 H.P., and which weighed altogether, engine and boiler combined, 80 lbs., and I applied its power in two separate experiments, the records of which will be found in the Reports of the Aeronautical Society, and also in the *Encyclopædia Britannica*. These experiments cost me many hundreds of pounds, and one of the experiments was witnessed by the Duke of Argyll, the Duke of Sutherland, Lord Dufferin, and others ; to whom I explained that a 30-H.P. engine would do 10 times the work of a 3-H.P. engine, but would not weigh more than five times 80 lbs. When this 3 H.P. engine lifted 120 lbs. before their eyes, I thought that the results warranted raising funds for further experiments, but I did not have the money myself, and I was unable to secure the assistance requisite to construct the larger machine.

On March 18, 1879, I constructed a model which, by its own power, traveled forward on the ground and rose a short distance from the ground. You will find this experiment recorded in the Report of the Aeronautical Society for that year, and I shall probably exhibit this model next year at one of the exhibitions.

I have sometimes been called sanguine, but I think this designation is a great mistake. I do not expect my countrymen to wake up to the importance of this subject during my lifetime. Fifteen years have passed since I gave the lesson of the albatross, and I expect it will take another 15 years before it is made use of.

Aerial navigation happens to be one of the subjects which I have studied, and it is a mistake to suppose that I am more anxious to effect that object than I am to effect improvements in marine propulsion and steam boilers and engines and other kindred subjects. Having quietly and patiently studied all these subjects, I now propose to give you a very simple and practical lesson on the flight of the albatross and its possible imitation by man.

You know, of course, that the albatross is the king of the sea, whose paradoxical soaring on motionless wings amazes travelers in southern seas. There are several varieties, but the great albatross is seldom found north of the 35th parallel of south latitude, and Mr. Wenham, in his celebrated paper on aerial locomotion, alluded to the bird as follows :

" One of the most perfect natural examples of easy and long-sustained flight is the wandering albatross"—a bird for endurance of flight probably unrivalled. Found over all parts of the Southern Ocean, it seldom rests on the water. During storms, even the most terrific, it is seen now dashing through the whirling clouds, and now serenely floating, without the least observable motion of its outstretched pinions. " The wings of this bird extend 14 or 15 ft. from end to end, and measure only 8½ in. across the broadest part. This conformation gives the bird such an extraordinary sustaining power, that it is said to *sleep* on the wing during stormy weather, when rest on the ocean is impossible. Rising high in the air, it skims slowly down, with absolutely motionless wings, till a near approach to the waves awakens it, when it rises again for another rest."

These are the observed facts, concerning which there is no dispute. The bird can remain indefinitely afloat upon the wind, with no muscular exertion save the infinitesimal effort required for steering and balancing himself, and the question arises as to how he utilizes the wind.

Those who understand how to handle a sailing-vessel are aware that by " tacking" you can work a vessel to windward. The operation is effected by what is called " close hauling ;" and any well-built fore-and-aft rigged vessel can sail within four points of the wind. If the line *A B* in diagram Fig. 84, indicates the direction of the wind, and the line *C D* the angle of the sail, then the line *E F* represents the actual force propelling the vessel forward. I presume that there is no one who will question the correctness of this. The wind may be blowing southeast, and the vessel can travel due east.

Now we come to the bird. I shall not go into the subject of the anatomy of the bird, or its food, etc., but would hope that there is no one here who holds that exploded theory that a bird

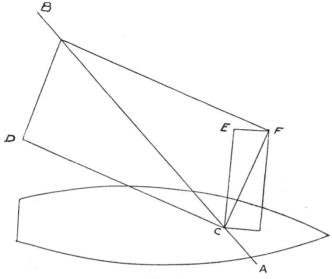

Fig. 84.

fills its quills with a few cubic inches of gas and thereby becomes lighter than the air.   Those bipeds, who shoot sea-gulls
or other flying creatures, know full well that when the birds
are maimed they fall with a thud ; and under no circumstances
can a bird materially diminish his own weight.

The albatross, when sailing over a very calm sea, is obliged
to flap its wings in order to keep up its speed.   But when a
strong wind is blowing it obtains the very impulse that it needs
without the necessity of using its wings as propellers ; the
waves are produced by the pressure of the wind upon the
water, and the weather face of the wave throws the wind upward, and gives the bird an upward thrust and impulse, of
which it takes advantage, and thus saves itself the exertion of
propulsion by flapping.   The vertical direction of the wind
above the wave-tops enables the bird to " close haul" and
thereby obtain gratuitous propulsion, which supplies his need
until he reaches the next opportunity.

Now to prove the existence of this upward current.   You
have seen the spray thrown up from the tops of the waves, you
have perhaps had a dash of it in your face while standing by
the weather gunwale.   A dash of spray over the bows of a
cutter yacht has made you feel rather damp.   These effects,
and others I might name, all prove the existence of this impulse
which serves the bird's wants, like the impulse imparted to a
pendulum at the commencement of its oscillation.

Now if you look at my second diagram, Fig. 85, you will
see what I mean.   The line *A B* represents here the upward

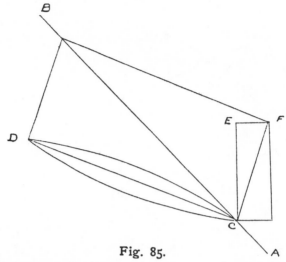

Fig. 85.

direction of the current, caused by the shape of the weather side of the wave, and you will at once perceive the similarity of the forces, but you must bear in mind that one is horizontal and the other vertical.

It is quite immaterial which way the bird travels, except that its speed is much greater when it sails with the wind, or even "on a wind"—that is, with a beam wind. It can go dead to windward, but it can also go in any other direction, just as a vessel can, not only go to windward, but also with the wind free. There are, however, several advantages possessed by the bird over the ship which time will not permit of noticing.

You will therefore understand that the bird literally lies on the upward current : takes a propulsive thrust by depressing its head and sails to the next wave apex.

I must, however, warn you that this is only one kind of flight, and it is this one kind of flight which I shall now show you offers great facilities for man's imitation ; and I now propose it as a first step toward conquering the air, not with cumbrous gas-bags and feeble screws rotated by leaky accumulators and a table-cloth for a rudder, but by following nature's perfect examples.

The method of flying which I have been attempting to describe to you is one which may be mastered by any intelligent man in a short time, and it is one which may be learned, in my judgment, with quite as much ease as bicycle riding.

His equipment will be a Boynton dress, a pair of stiff, immovable wings or aeroplanes, a light sea-anchor with a length of rope attached, and an accompanying boat. The frame of the planes will be extended behind so as to carry a horizontal rudder.

The Boynton dress and the aeroplanes will be firmly attached together. Each plane might be 18 ft. × 3 ft., and made stiff enough to bear three times the pressure that the total weight would impose upon them. If the total weight of man and equipment amounted to 250 lbs., I should test the strength of the planes by resting the center on its back and spreading upon the wings bags of sand to the amount of 750 lbs. If it sustained this weight fairly, then it would be safe for an experiment.

The albatross sometimes gets its initial velocity by starting from the apex of a wave while floating. In like manner the man could get his start by floating on the wave-top and using his sea-anchor, composed of several small hoops covered with canvas and open at one end. Holding on to the anchor and slacking out slowly, with his head to the wind, he would watch his opportunity, and pull the rope at the moment that he arrived at the top of a wave, when the wind would lift him just as a boy's kite is raised, and he could then go forward and pick up his anchor, emptying out the water as he did so. Every time he approached the apex of a wave (say from 10 to 20 ft. above it) he would depress the rudder, the effect of which would be to place his body momentarily in the position shown in diagram Fig. 85. This, with the impulse of the wind, would give him a high velocity which would not be very much lessened before he received another impulse.

Suppose a wind of 40 miles an hour were blowing, this line $E\ F$ would represent a propulsive force of about 50 lbs.; but the forward movement would probably reduce th s to 30 lbs., and this would be an ample propelling force for the purpose.

If the initial velocity were obtained by the same means that birds obtain it, from a moderate elevation, it would be quite easy to skim over the water at 10 or 20 ft. above the waves.

The first experiment would, of course, be in company with a boat to render assistance in obtaining the first rise from the water, and in any other emergency. After a few experiments a mile or two would be easily covered, and after some practice the speed would far exceed that of the bird, because the weight is so much greater.

With a well-inflated Boynton dress and a good sea-boat in attendance there would be no danger to life.

If you ask of what use is this? I say business, pleasure, healthful exercise, and the accomplishment of the initial step toward actual flight.

If a man thus equipped can go in the teeth of a gale and carry a rope to a wreck, that is business.

If a man thus equipped can travel at from 20 to 40 miles an hour over the waves, that is pleasure and healthful exercise.

But it will also teach him the enormous sustaining power to be derived from swift motion, and it will also explode many of the silly, unmechanical notions which are now held upon the subject.

We have had 100 years of balloons, and it is quite time that some advance should be made, leaving drifting bladders behind.

THE FLYING MAN.

THE CARRYING CAPACITY OF ARCHED SURFACES IN SAILING FLIGHT.*

By Otto Lilienthal.

[After the foregoing pages were all in type, the index printed and ready to go to the binder, a paper was received from Herr Lilienthal describing his 1893 experiments, which so fully sustains the views set forth in this book, and holds out such promise of success in the near future, that it has been decided, at some inconvenience, to include a translation of this paper in an extra appendix.]

I present herewith an account of my personal experience in soaring flight during the past year, this being the third annual report published by me in this journal. I have now reached the close of a series of experiments during which I had set myself a definite task. This was to construct an apparatus with curved carrying surfaces which should enable me to sail through the air, starting from high points and gliding as far as possible—that is to say, at the least obtainable inclination ; and to do this with stability and safety even in winds of medium strength.

I may be permitted, in giving this account of experiments in which artificial arched wings were used, perhaps for the first time, in a certain form of flight, to refer again to the introduction of this important element in the *technics* of flight.

Four years ago I completed a series of experiments on the carrying capacity of arched wings, which I had carried on with my brother for many years, and I published the results in my book, " The Flight of Birds as the Basis of the Art of Flying." In this work a new theory of flying, fairly explaining all the phenomena of bird-flight, was for the first time completely made known. The novel data involved in our calculations, and the remarkable results arrived at in our experiments, were my motives for not coming sooner before the public, until by numerous and repeated trials every chance of error appeared to be eliminated.

Now I have been accused in No. 6 of last year's issues of this journal* with having done wrong in keeping back these discoveries so long. I am told that I should have come before the public with them immediately after my first discovery of the new laws of air resistance more than 20 years ago. That while I was censuring the balloon for a delay of many decades, I was myself postponing the solution of the question of aerial navigation for two decades more by my silence.

* Translated from *Zeitschrift für Luftschiffahrt und Physic de Atmosphare* for November, 1893.

Flattering as it is for me to see so great an influence on the solution of the problem of flying attributed to the new discoveries presented by me, yet I will not forbear to point out that by these researches the theory of flight was merely put upon a somewhat sounder basis, but that for the full solution of the problem other very important questions will have to be settled.

The mere discovery that with arched wings supporting forces are evolved which permit soaring to be performed with little effort is far from being the final invention of flying. The successful practical utilization of this important phenomenon in air resistance is going to demand a considerable amount of ingenuity. To get the upper hand over the wind with flying machines and to bring about a beneficial utilization of those favorable supporting forces—for such a task many a technical man

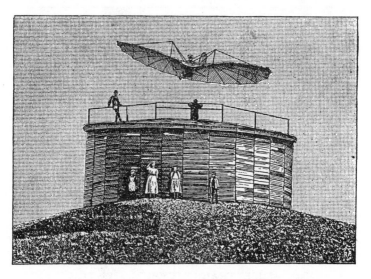

FIG. I.—LILIENTHAL'S "JUMPING-OFF" PLACE.

will have a chance to throw his talent into the scale ; for the field of work lying before us is no small one.

The great delay in publishing discoveries of myself and brother on aviation was only the natural result of the circumstances under which they were made. Even when we were devoting every hour of our leisure to the question of aviation, and were already on the track of the laws which were to evolve this problem, people in Germany still considered every man who occupied himself with this unprofitable art as little better than a lunatic. This was sufficient cause for our not attracting unnecessary attention to such studies. The principal professor of mathematics at the Berlin Gewerbe Academie in the sixties heard from one of my fellow-students—they had given me even then an appropriate nickname—that I was occupied with investigations in aeronautics. The professor sent word to me that there was no harm in my amusing myself with such calculations, but that I "should, for Heaven's sake, not spend any money for such things." The professor, it is true, did not know that this last advice was (for cause) quite unnecessary.

At that time a special learned commission appointed by the State had just *officially* declared, once for all, that man would never be able to fly ; by which declaration naturally public opinion, in favor of experiments on this problem, was not exactly stimulated. Even later on the interest in aviation was much less than it is to-day. German societies for the advancement of aerial navigation were not yet in existence ; and when

FIG. 2.—LILIENTHAL'S "JUMPING-OFF" PLACE.

they were formed the balloon, which I have always considered an obstacle in the development of free flight, monopolized their entire attention. For this reason I did not at that time think it advisable to join any one of the societies. When, however, I had done so, and aviation had other representatives in the society, I soon took advantage of the opportunity by communicating our experimental results in a series of lectures, which were followed shortly by the publication of my book.

To-day, with clear and comprehensive diagrams before us, the explanation of the flight of birds seems very simple and natural, while formerly every crow that flew by assigned us the puzzle of its slowly moving wings to solve. To-day, too, it is easy, in investigations on air resistance, to take surfaces curved like the wings of a bird instead of flat plates, and to develop in succession those wonderful effects, the first discovery of which was, after all, not quite so simple and self-evident as some may be inclined to assume now. The thoughtful reader will readily understand that it took much time and study to arrive at the conclusion that the slight curvature of the wing was the real secret of flying. This was a conclusion which we reached purely as a logical deduction from the flight of birds, and which some noted investigators are not willing to admit even at this late day. Besides this, being young men entirely without resources, we were compelled to procure the means for carrying on our experiments by the most petty economies, and at times to suspend entirely our work in aeronautics by reason of the struggle for existence. We should not at that time in any event have been able to publish our results in proper form. Even at the last I had the very greatest difficulty in finding a publisher for my book.

Although, then, the delay in publishing my work was simply the result of circumstances, the position assumed toward it, in the sequel, by certain investigators, proved to me that my satisfaction in my work would have been much diminished if it had been published before the entire data were on hand in a complete form.

I must confess, however, that to us who abandoned *flat* wings fully two decades ago, it seemed almost inconceivable that experimenters should cling so tenaciously to the aeroplane and to hopeless calculations on the resistances of plane surfaces, as practically all of them did during this whole period. Even to the present hour the majority of aviators expend much painful effort in attempting the hopeless task of trying to fly with flat wings.

Our expectation that the publication of our experiments, showing the problem of flight in an entirely new light, would be followed by practical tests as to the correctness of the new laws indicated by us was not at once realized. Our results stood alone and unsupported until very lately, and have but recently been corroborated by the work of Wellner and Phillips. Even now the flat wing does not show any signs of disappearing from the field very soon.

When it became known that the French Government* had granted funds for experiments on air resistance, I sought to induce the president of the commission in charge of them to include curved surfaces as well as flat plates within the field of his experiments. At the same time I sent him a copy of my work, and by means of my diagrams pointed out to him the favorable results in air resistance obtained with curved wings.

M. Hureau de Villeneuve thanked me for the book, pronounced it to be very interesting, but declined to grant my request as to the consideration of hollow surfaces in his researches.

His reasons for this course literally translated were the following :

---

* He means the French Society for the Advancement of Science.—ED.

FIG. 3.—LILIENTHAL'S FLIGHT.

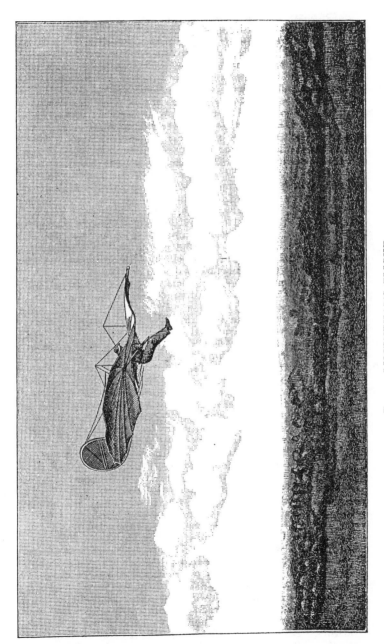

FIG. 4.—LILIENTHAL'S FLIGHT.

" I consider wings, pre-eminently, as moving organs for propelling at great speed, and when this speed has been reached
the force necessary for *supporting* may be neglected. I have
shown this by constructing mechanical birds with flat wings,
which fly very fast and keep themselves up very well when
flying in the air. I try to make the wings true planes as far as
I can, for the flatter they are the greater is the speed attained."
After this expression of opinion there is, I think, for the
present little prospects that experiments to be made in aeronautics with the aid of the French Government will be of any considerable benefit to the technics of flight. It is also to be
greatly regretted that the extended experiments of Professor
Langley, carried out with so much care and expense, were limited to the resistance of plane surfaces, as for such surfaces
data have long been on record sufficient for computing all cases
occurring in practice.

German and Austrian aviators, it is true, since the publication of my book have largely put aside flat wing and sail surfaces and introduced arched wings. However, this was done
mainly on paper, in projects, and in aeronautical papers and
discussions. I therefore felt impelled all the more to carry out
myself my theory in practice, and to utilize the results of my
preliminary small scale experiments. Through my long famil-
iarity with air and wind I had come to the conclusion that a
particular class of difficulties was next to be overcome. In
trials with movable wings, in the building of steam air ships,
in experiments with mechanical birds of all kinds, I had found
out how hard it is to maintain a *stable* equilibrium in the air
and to counteract the " whims" of the wind.

For this reason I gave up for the time being motor mechan-
isms altogether, and limited myself to the simplest form of
flight—namely, gliding downward in an inclined direction.
The object of this was to ascertain practically whether *stable*
sailing and equilibrium through the air was possible ; whether
by practice we could learn to overcome the wind, that arch-
enemy of all wing builders, and, above all, whether with large
" human" wings the favorable carrying capacities of arched
wings would be maintained, which our experiments in small
wings (not over ½ sq. m. in area) had shown.

It was further my purpose to extend my former experience
in practical wing construction, and to build safer and lighter
sailing apparatus which should permit soaring safely at considerable heights. As a final result we would also ascertain at
what angles of inclination sailing flight can be maintained with
and without the wind, and what conclusions can be drawn from
this as to the " work" or power required for flying in general.

In addition to the details given in my communications of
1891 and 1892, I will mention briefly that this spring I erected
a special flying station on the " Mai" hill, near Steglitz. The
owner of the property, Mr. Seldis, allowed me in the kindest
way to arrange a little earthmound at this spot for experiments
in soaring. By building a shed in the shape of a tower, from
the roof of which I sailed, I obtained a " jumping off" place

10 m. (33 ft.) high. The shed serves also for storing my apparatus. The earth slope around the shed falls off toward the southwest, the west, and the northwest. The roof is covered with sod 'to give a firm footing in taking a run, and slopes in the above-mentioned directions. In the assumption that I had thus provided sufficiently for the frequent changes in the direction of the wind I was entirely mistaken. During the first half of this summer the wind varied almost entirely between the east and the north, and my flying tower remained unused for nearly three months. The wind, be it understood once for all, cannot be depended upon. A hill well suited for trials in sailing flight should consist of a cone sloping on all sides, so that the start may be made against the wind in any direction.

The two figures (figs. 1 and 2) show the " jumping-off ' place from this tower, from in front as well as from the side. The second figure shows also the attachment of the apparatus to the body, already described in the past, which consists in simply grasping it with both hands.

A novelty that I introduced this year in my apparatus is the possibility of folding it up. The wings are formed of ribs arranged radially and can be closed up, somewhat like the wings of a bat. In this way it became more portable and can be stored anywhere.

I have gradually given up the great "spread" of my former apparatus. The unequal strength and direction of the wind, under the right or the left wing, frequently produce a considerable displacement of the center of action of the air pressure, which becomes worse the greater the " spread" of the wings. This spread I now never make greater than 7 m. (23 ft.), and I am thus always able to restore equilibrium by a simple change in my center of gravity.

The breadth of the wings has also its limits. It must be possible, in an instant, to transfer the center of gravity so far from front to back as the point of action of the supporting air resistance can move. The furthest backward limit of this motion is the center of gravity of the wing surface. When you fall vertically with this apparatus in calm air it acts entirely as a parachute. The air strikes vertically from below and presses uniformly on all parts of the surface. It is possible to get into this position in flying if the aviator uses up, in gliding upward, all the *vis-viva* acquired in flying forward and downward, whence the velocity will decrease very rapidly on the upward turn. I have often been obliged thus to pass over some obstacles—a tree, a crowd of people, or the photographer who was getting a view of me from the front. It is easy to rise in this way, but the apparatus comes to a stand at the summit, and if it cannot be properly and promptly slanted down behind, it will be tilted toward the front in falling, and the front edge will be broken when it touches the ground.

The center of the supporting air pressure reaches its limit *toward the front,* when the machine is struck by a stronger wind gust in flying. It is then necessary for the aviator to throw himself as far forward as possible and also to stretch his legs forward, otherwise velocity is lost, and the machine will be driven back by the wind. The figure (No. 3) shows such a critical position. I had started in a wind estimated by me at from 6 to 7 m. velocity per second (13 to 16 miles per hour), and while I was in motion the velocity of the air rose to probably about 10 m. (22 miles), for I was suddenly raised with so much force that I became suspended in the air at a higher level than my point of starting. Then I remained fixed at this point for several seconds until the wind decreased, when I again sailed forward and moved downward slowly. In order to be able to conveniently equalize this fore-and-aft motion of air pressure, by a change in the center of gravity, it is advisable to make the breadth of wing not greater than 2½ m. (8.2 ft.). From these conditions we arrived at a pretty definite relation in the shape and the size of the wings. With a spread of 7 m. (23 ft.) and a breadth of 2½ m. (8.2 ft.) we get, allowing for the rounding of the ends, an area of 14 sq. m. (151 sq. ft), which is sufficient to carry the mean weight of a man. Such wings weigh about 20 kilograms (44 lbs.). My own weight is 80 kilograms (176 lbs.), giving a total weight of exactly 100 kilograms (220 lbs.).

I have now moved my main practice ground to the Rhinow Hills, which I referred to in my last year's communication. The topography is here as if made especially for experiments in flying. Out of the surrounding flat fields a range of hills, covered only with grass and heather, rises to a height of 60 m. (200 ft.) with inclined slopes on all sides. The inclinations of the slopes vary from 10° to 20°, so that one can select a place for sailing along in any direction. When, this year, I unfolded my flying outfit for the first time on these slopes, a somewhat anxious feeling came over me, when I said to myself, "From up here you are now to sail down into the land spread out deep before you." But the first few careful leaps soon gave me back a feeling of security, as the soaring flight began here far more gently than from my flying tower. The wind did not "rear up" here as it did in front of the tower, where I always felt a sudden upward blast from the wind in passing the edge, which often threatened to be dangerous.

You run down hill against the wind with lowered wings, at the proper moment you raise the carrying surface up a little, so that it is approximately level ; and then, springing forward, you try, by a proper position of the center of gravity of the apparatus, to give it such an inclination that it will glide along rapidly and drop as little as possible. Beginners will do well to select for this a gentle slope, over which they may glide along a low elevation. The first rule is to keep your legs well extended toward the front, and in landing to throw the upper part of the body backward, so that the front edge raises itself and thus checks the motion, as may be seen whenever a crow alights. The starting and landing must be done exactly square against the wind. The fixed vertical rudder will keep the apparatus exactly in the wind when in a state of rest. The horizontal rudder keeps the apparatus from tipping over forward, a thing that arched surfaces are inclined to .do. In landing, however, the horizontal rudder must not hinder a rapid tipping up of the machine ; hence it must turn freely on its forward edge when the air presses upward, and its motion must be limited at the bottom only.

The following mistake is to be particularly avoided. The experimenter is soaring in the air and feels himself suddenly raised by the wind, but unequally, as is usually the case—for instance, the left wing more than the right. The inclined position forces him toward the right. The beginner involuntarily stretches his legs to the right, because he foresees that he will strike the earth on the right hand. The result is that the right wing, which is already lower, is loaded still more, and flight tends more and more downward and to the right, until the tips of the right wing strike the earth and are broken. For life and limb there is less danger, as the apparatus forms an efficient guard in every direction, which checks the force of the blow. The correct thing to do is always to extend one's legs toward the wind that is rising, and thus to press it down again. In the beginning this requires some force of will, but this useful movement soon becomes an unconscious one, after we see how surely the wings can be guided this way and be protected from damage.

Beginners are also easily tempted to utilize the momentum which the easy downward glide gives them for a bold upward flight. They are likely to forget, however, that at the summit of the rising curve their apparatus becomes a mere parachute. They do not lean back far enough and incur damages requiring considerable repairs to the machine. It is better to postpone such tricks until the regulating of the center of gravity has become a second nature, like balancing on the bicycle, and rather to lay most stress on gliding ahead at an inclination which shall be uniform and as slight as possible, and particularly to endeavor to make an elegant landing. Of course we do not have time enough when in the air to consider carefully whether the position of the wings at any particular time is the correct one. This is entirely a matter of practice and experience. After a few trials one begins to have a feeling of mastery over the situation. A consciousness of safety crowds out the first feeling of anxiety. Finally we become perfectly at ease, even when soaring high in the air, while the indescribably beautiful and gentle gliding over the long sunny slope rekindles our ardor anew at every trial. It does not take very long before it is quite a matter of indifference whether we are gliding along 2 m. or 20 m. (6 ft. or 65 ft.) above the ground ; we feel how safely the air is carrying us, even though we see diminutive men looking up at us with astonishment. Soon we pass over ravines as high as houses, and sail for several hundred meters through the air without any danger, parrying the force of the wind at every movement.

After we have reached a feeling of safety in direct flight against the wind, we involuntarily try, first a very little and then more and more, to direct the course of . our flight to the right and to the left. A slight change in the center of gravity to one side produces at once a small inclination of the carrying surface, so that the supporting air pressure also moves to this side and changes the direction of flight. There is nothing simpler than steering flying machines It is important not to forget in this connection that the landing must always be made dead against the wind.

On the occasion shown by fig. 4, which represents a very long and high flight, I had carried the deviation from straight flight so far that at times I was flying almost in an opposite direction. Coming from the hill at the right hand, I had almost turned my back to the plain at the moment the picture was taken.

For the present it does not appear to me to be advisable to endeavor to bring the body into a horizontal position, because the legs must always be ready for action in running, jumping, steering and landing. Later on it will, perhaps, be possible to make this change, after some other impòrtant improvements have been made in the apparatus. Of course this would be very important for producing easier passage through the air and the consequent saving of energy in flying.

In the annexed figure (5) several lines of flight are illustrated. The dotted line *d e* shows the path traversed in a calm. At the top of the hill, at the point *a*, a running start is taken with as high speed as possible and with wings lowered. The steeper the hill the better. Then at *b* you raise your wings a very little in front and try to glide along as close to the ground as possible, with your legs extended to the front. The diagram at *C* shows how the air resistance *L* has at the same time a supporting and propelling force or component. Soon our velocity has increased so much that at *d* we can change to a flatter inclination. Thus at an angle of 9° to 10° we approach the valley and do not raise our wings in front till the very end at *e*, when we apply a strong force by throwing the upper part of our body back. Then the pressure of the air, acting suddenly at the front, checks the motion, and we drop on to our feet quite gently, just as if we had jumped, without wings, from the height of a chair.

When there is any wind the motion is still more gentle. Flight against the wind, in reference to the earth, is in all cases slower, as, of course, the *relative* motion of the air and the apparatus governs the speed. The relative velocity attained is felt by the strength of the breeze striking the face while flying. A convenient device would be a little indicator pressure gauge in the front of the apparatus, on which we could constantly read the relative speed of the air. This would not involve any appreciable increase in the resistance.

Although the wind compels us to resort to various extraordinary manœuvres, it also furnishes us an opportunity for testing the real value and scope of sailing flight. By our calculations, based on experiments with arched surfaces on a small scale during windy weather, extended and prolonged sailing flight can be explained without further trouble. With wings of

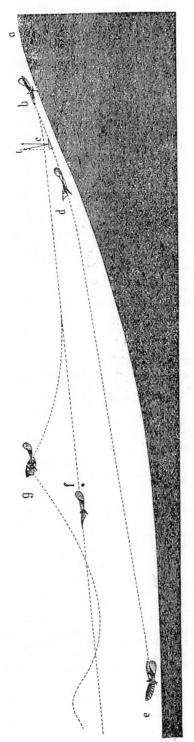

Fig. 5.—LINES OF LILIENTHAL'S FLIGHT.

a proper form and position, the wind needs only to reach the necessary strength in order to keep the experimenter from falling. Even with light winds of 4–5 m. velocity per second (9 to 11 miles per hour), we can with some little practice glide along at the slight angle of 6° to 8°, as is shown in the line *b f*.

The greatest velocity of the wind at which I dared to start was about 7–8 m. per second (15 to 18 miles per hour). In these flights I often had a very interesting though not dangerous struggle with the wind, in which I sometimes came to a state of absolute rest, and was suspended in the air at one point for several seconds, almost exactly as the falcons of the Rhinow Mountains are. Sometimes I was suddenly lifted from such a position of rest many meters in a vertical direction, so that I became alarmed lest the wind should carry me off altogether. As, however, I never ventured out except when such gusts were exceptional, I was always able to continue my flight and to land safely. The line *b g* shows a wavy course, brought about by gusts, during which I rose to the height of my point of starting.

I want to emphasize particularly that the results obtained in actual sailing agree well with my small-scale experiments on the supporting power of arched surfaces. During a calm it is quite feasible, with proper practice in placing the wings, to sail downward at an angle of 9°. Observation shows that the wings are then approximately level. According to my diagrams on Plate VII* the supporting force of the air would then be 80 per cent. of the resistance of the surfaces normally acted upon. In a calm the velocity of flight as measured was about 9 m. per second, with a surface of 14 sq. m.; the uplifting force would be $= 0.8 \times 0.13 \times 14 \times 9^2 = 118$ kilograms, while 100 kilograms would suffice.

According to the diagram on Plate VI, if a curved surface is struck at an angle of 9°, its *pressure* will act toward the front at an angle of 3° with the normal to the chord of the surface. The propelling effect of the air pressure will therefore be :

$$100 \times sin. \ 3° = 5 \ \text{kilograms.}$$

This force is used up in propelling through the air the cross-sections of the body and of the framework at a velocity of 9 m. per second. This would correspond to a projected area F, to be deduced from the equation :

$$5 \ \text{kgm.} = 0.13 \times F \times 9^2,$$

which gives $F = 0.5$ sq. m., a result agreeing well with the actual fact.

In a medium wind a practised "sailer" can pass over considerable space at an angle of 6°, as I have proved from experience ; the velocity of motion in this case is about 5 m. per second. It is a fair assumption that the wind as it strikes the mountains has a greater ascending trend, but I succeeded in maintaining the same angle of descent on somewhat narrow

---

* Lilienthal : "Der Vogelflug als Grundlage der Fliegekunst."

hill-sides, where the wind could escape easily to either side, and also at a considerable distance away from the mountains. For this reason I think it likely that the same results could be obtained upon extended flights.

There can be no doubt, in my opinion, that by perfecting our present apparatus, and by acquiring greater skill in using it, we shall achieve still more favorable results with it, and finally succeed in taking long sails even in rather strong winds. Even without considering the chances of such continued sailing without effort, the results already obtained provide us with data as to the energy to be expended if horizontal flight is to be prolonged by mechanical means.

The wind seldom has a velocity exceeding 4–5 m. per second. Sailing against the wind there will be a drop per second of :

$$5 \sin. 6° = 0.5 \text{ m.} = 1.64 \text{ ft.}$$

If we disregard minor details the "work" of my apparatus will, therefore, be : $100 \times 0.5 = 50$ kilogrammeters per second, or $\frac{2}{3}$ H.P. This is for the purely ideal case, when the wings move downward uniformly in the line of flight, and are raised without appreciable loss of time ; but the weight of the motor, and the decrease in carrying force during the upward stroke of the wings, will, of course, increase the total work necessary. I have already completed a steam-engine which I intend to use for moving wings in some experiments to be made in the near future. It develops 2 H.P., will work half an hour, and weighs 20 kilograms (44 lbs.), including all its accessories. In this way the "work" required will be increased to 60 kilogrammeters per second. When the separate surfaces at the tip of the wings, corresponding to the primary feathers, move up and down, 1st, the surfaces will move at a more favorable angle of incidence, and, 2d, a further increase in lifting power is produced by the beating action of the wings, both of which facts decrease the necessary expenditure of power. It thus appears possible that we may succeed in maintaining horizontal flight with the above-mentioned motor, though the upright position of the body is unfavorable to flying.

Of course it will be a matter of practice to learn how to guide such a flying machine, with beating wings, as surely as a simple sailing apparatus. For this reason I shall first use my new machine with rigid wings as a simple sailing apparatus. Later on, when I have again acquired perfect confidence, I shall allow the tips of the wings to make very small beats, and shall increase these very gradually to their full stroke.

In this way, passing by small gradations from sailing flight to rowing flight with wings, the length and duration of flight may be considerably increased, so that we may venture at last to fly *with* the wind for a time and then try flying in circles.

In my trials with such large carrying surfaces in flying I have found that the concavity of the wings should be less than results from my experiments in small models. While with small surfaces, of less area than 1 sq. m., a depth of one-twelfth of the breadth of the wing gave the best results, on the large wings of 14 sq. m., a depth of one-eighteenth to one-twentieth of the breadth proved to be the best. I have made frequent changes in the profile of the wings to make certain of this fact. My apparatus was specially arranged to permit this.

Of late much stress has been laid on the parabolic cross-section of the bird's wings. It is evident that the "projectile curve" would be the natural form for diverting the air current under the wing. Apparently the various profiles actually represented have been drawn rather arbitrarily. If we compare the circle with the half vertex of the parabola, which is the part generally shown, the difference is found to be so slight as to be hardly worth considering. On the other hand, in the sketches so presented by others, the front of the arch shows decidedly more curvature than the back; this was probably done to explain the greater inclination of the air pressure in front. Of course, it is quite feasible to give up the ordinary parabola of the second degree, and use instead a parabola of a higher order, which rises at first more steeply and runs out more gradually. During three years of practical experiments I have repeatedly tried such profiles in sailing, but I can only advise strongly against making the surface too steep in front, as there is a risk of getting a dangerous upward pressure. My advice is to stick to the ordinary parabola even when it almost coincides with a circle.

In closing, I would express the wish that the publication of my results may incite others to take up the problem of flight practically. I represent merely but one certain line of thought, which has led me so strongly to the imitation of the flying birds, that I have even been accused of "aping' them. Such accusations cannot, however, divert me from the path I have chosen. I do not imitate the flight of the birds because it happens to be convenient to copy, but because it combines logical correctness with so many practical advantages which no other principle of flying could furnish me.

The only method which, in my opinion, might to a certain degree offer the advantages of moving wings is that of rotating air propellers, which lift and propel at the same time. I have already expressed my opinion in the past that possibly good results may be obtained with such devices if skilfully worked out. Actual trial alone can decide this question, as we must let the air and the wind have their say in the matter. For this reason it would be desirable to have some such ideas carried out in practice, so that there will be an end to fruitless discussions.

GENERAL INDEX.

A CATALOG OF SELECTED
DOVER BOOKS
IN ALL FIELDS OF INTEREST

A CATALOG OF SELECTED DOVER
BOOKS IN ALL FIELDS OF INTEREST

CONCERNING THE SPIRITUAL IN ART, Wassily Kandinsky. Pioneering work by father of abstract art. Thoughts on color theory, nature of art. Analysis of earlier masters. 12 illustrations. 80pp. of text. 5⅜ x 8½. 23411-8 Pa. $3.95

ANIMALS: 1,419 Copyright-Free Illustrations of Mammals, Birds, Fish, Insects, etc., Jim Harter (ed.). Clear wood engravings present, in extremely lifelike poses, over 1,000 species of animals. One of the most extensive pictorial sourcebooks of its kind. Captions. Index. 284pp. 9 x 12. 23766-4 Pa. $12.95

CELTIC ART: The Methods of Construction, George Bain. Simple geometric techniques for making Celtic interlacements, spirals, Kells-type initials, animals, humans, etc. Over 500 illustrations. 160pp. 9 x 12. (USO) 22923-8 Pa. $9.95

AN ATLAS OF ANATOMY FOR ARTISTS, Fritz Schider. Most thorough reference work on art anatomy in the world. Hundreds of illustrations, including selections from works by Vesalius, Leonardo, Goya, Ingres, Michelangelo, others. 593 illustrations. 192pp. 7⅛ x 10¼. 20241-0 Pa. $9.95

CELTIC HAND STROKE-BY-STROKE (Irish Half-Uncial from "The Book of Kells"): An Arthur Baker Calligraphy Manual, Arthur Baker. Complete guide to creating each letter of the alphabet in distinctive Celtic manner. Covers hand position, strokes, pens, inks, paper, more. Illustrated. 48pp. 8¼ x 11. 24336-2 Pa. $3.95

EASY ORIGAMI, John Montroll. Charming collection of 32 projects (hat, cup, pelican, piano, swan, many more) specially designed for the novice origami hobbyist. Clearly illustrated easy-to-follow instructions insure that even beginning papercrafters will achieve successful results. 48pp. 8¼ x 11. 27298-2 Pa. $3.50

THE COMPLETE BOOK OF BIRDHOUSE CONSTRUCTION FOR WOOD-WORKERS, Scott D. Campbell. Detailed instructions, illustrations, tables. Also data on bird habitat and instinct patterns. Bibliography. 3 tables. 63 illustrations in 15 figures. 48pp. 5¼ x 8½. 24407-5 Pa. $2.50

BLOOMINGDALE'S ILLUSTRATED 1886 CATALOG: Fashions, Dry Goods and Housewares, Bloomingdale Brothers. Famed merchants' extremely rare catalog depicting about 1,700 products: clothing, housewares, firearms, dry goods, jewelry, more. Invaluable for dating, identifying vintage items. Also, copyright-free graphics for artists, designers. Co-published with Henry Ford Museum & Greenfield Village. 160pp. 8¼ x 11. 25780-0 Pa. $10.95

HISTORIC COSTUME IN PICTURES, Braun & Schneider. Over 1,450 costumed figures in clearly detailed engravings–from dawn of civilization to end of 19th century. Captions. Many folk costumes. 256pp. 8⅜ x 11¾. 23150-X Pa. $12.95

STICKLEY CRAFTSMAN FURNITURE CATALOGS, Gustav Stickley and L. & J. G. Stickley. Beautiful, functional furniture in two authentic catalogs from 1910. 594 illustrations, including 277 photos, show settles, rockers, armchairs, reclining chairs, bookcases, desks, tables. 183pp. 6½ x 9¼.   23838-5 Pa. $9.95

AMERICAN LOCOMOTIVES IN HISTORIC PHOTOGRAPHS: 1858 to 1949, Ron Ziel (ed.). A rare collection of 126 meticulously detailed official photographs, called "builder portraits," of American locomotives that majestically chronicle the rise of steam locomotive power in America. Introduction. Detailed captions. xi + 129pp. 9 x 12.   27393-8 Pa. $12.95

AMERICA'S LIGHTHOUSES: An Illustrated History, Francis Ross Holland, Jr. Delightfully written, profusely illustrated fact-filled survey of over 200 American light-houses since 1716. History, anecdotes, technological advances, more. 240pp. 8 x 10¾.
25576-X Pa. $12.95

TOWARDS A NEW ARCHITECTURE, Le Corbusier. Pioneering manifesto by founder of "International School." Technical and aesthetic theories, views of industry, economics, relation of form to function, "mass-production split" and much more. Profusely illustrated. 320pp. 6⅛ x 9¼. (USO)   25023-7 Pa. $9.95

HOW THE OTHER HALF LIVES, Jacob Riis. Famous journalistic record, exposing poverty and degradation of New York slums around 1900, by major social reformer. 100 striking and influential photographs. 233pp. 10 x 7⅞.
22012-5 Pa. $10.95

FRUIT KEY AND TWIG KEY TO TREES AND SHRUBS, William M. Harlow. One of the handiest and most widely used identification aids. Fruit key covers 120 deciduous and evergreen species; twig key 160 deciduous species. Easily used. Over 300 photographs. 126pp. 5⅜ x 8½.   20511-8 Pa. $3.95

COMMON BIRD SONGS, Dr. Donald J. Borror. Songs of 60 most common U.S. birds: robins, sparrows, cardinals, bluejays, finches, more–arranged in order of increasing complexity. Up to 9 variations of songs of each species.
Cassette and manual 99911-4 $8.95

ORCHIDS AS HOUSE PLANTS, Rebecca Tyson Northen. Grow cattleyas and many other kinds of orchids–in a window, in a case, or under artificial light. 63 illustrations. 148pp. 5⅜ x 8½.   23261-1 Pa. $4.95

MONSTER MAZES, Dave Phillips. Masterful mazes at four levels of difficulty. Avoid deadly perils and evil creatures to find magical treasures. Solutions for all 32 exciting illustrated puzzles. 48pp. 8¼ x 11.   26005-4 Pa. $2.95

MOZART'S DON GIOVANNI (DOVER OPERA LIBRETTO SERIES), Wolfgang Amadeus Mozart. Introduced and translated by Ellen H. Bleiler. Standard Italian libretto, with complete English translation. Convenient and thoroughly portable–an ideal companion for reading along with a recording or the performance itself. Introduction. List of characters. Plot summary. 121pp. 5¼ x 8½.
24944-1 Pa. $2.95

TECHNICAL MANUAL AND DICTIONARY OF CLASSICAL BALLET, Gail Grant. Defines, explains, comments on steps, movements, poses and concepts. 15-page pictorial section. Basic book for student, viewer. 127pp. 5⅜ x 8½.
21843-0 Pa. $4.95

▲

BRASS INSTRUMENTS: Their History and Development, Anthony Baines. Authoritative, updated survey of the evolution of trumpets, trombones, bugles, cornets, French horns, tubas and other brass wind instruments. Over 140 illustrations and 48 music examples. Corrected and updated by author. New preface. Bibliography. 320pp. 5⅜ x 8½. 27574-4 Pa. $9.95

HOLLYWOOD GLAMOR PORTRAITS, John Kobal (ed.). 145 photos from 1926-49. Harlow, Gable, Bogart, Bacall; 94 stars in all. Full background on photographers, technical aspects. 160pp. 8⅜ x 11¼. 23352-9 Pa. $12.95

MAX AND MORITZ, Wilhelm Busch. Great humor classic in both German and English. Also 10 other works: "Cat and Mouse," "Plisch and Plumm," etc. 216pp. 5⅜ x 8½. 20181-3 Pa. $6.95

THE RAVEN AND OTHER FAVORITE POEMS, Edgar Allan Poe. Over 40 of the author's most memorable poems: "The Bells," "Ulalume," "Israfel," "To Helen," "The Conqueror Worm," "Eldorado," "Annabel Lee," many more. Alphabetic lists of titles and first lines. 64pp. 5³⁄₁₆ x 8¼. 26685-0 Pa. $1.00

PERSONAL MEMOIRS OF U. S. GRANT, Ulysses Simpson Grant. Intelligent, deeply moving firsthand account of Civil War campaigns, considered by many the finest military memoirs ever written. Includes letters, historic photographs, maps and more. 528pp. 6⅛ x 9¼. 28587-1 Pa. $11.95

AMULETS AND SUPERSTITIONS, E. A. Wallis Budge. Comprehensive discourse on origin, powers of amulets in many ancient cultures: Arab, Persian Babylonian, Assyrian, Egyptian, Gnostic, Hebrew, Phoenician, Syriac, etc. Covers cross, swastika, crucifix, seals, rings, stones, etc. 584pp. 5⅜ x 8½. 23573-4 Pa. $12.95

RUSSIAN STORIES/PYCCKNE PACCKA3bl: A Dual-Language Book, edited by Gleb Struve. Twelve tales by such masters as Chekhov, Tolstoy, Dostoevsky, Pushkin, others. Excellent word-for-word English translations on facing pages, plus teaching and study aids, Russian/English vocabulary, biographical/critical introductions, more. 416pp. 5⅜ x 8½. 26244-8 Pa. $8.95

PHILADELPHIA THEN AND NOW: 60 Sites Photographed in the Past and Present, Kenneth Finkel and Susan Oyama. Rare photographs of City Hall, Logan Square, Independence Hall, Betsy Ross House, other landmarks juxtaposed with contemporary views. Captures changing face of historic city. Introduction. Captions. 128pp. 8¼ x 11. 25790-8 Pa. $9.95

AIA ARCHITECTURAL GUIDE TO NASSAU AND SUFFOLK COUNTIES, LONG ISLAND, The American Institute of Architects, Long Island Chapter, and the Society for the Preservation of Long Island Antiquities. Comprehensive, well-researched and generously illustrated volume brings to life over three centuries of Long Island's great architectural heritage. More than 240 photographs with authoritative, extensively detailed captions. 176pp. 8¼ x 11. 26946-9 Pa. $14.95

NORTH AMERICAN INDIAN LIFE: Customs and Traditions of 23 Tribes, Elsie Clews Parsons (ed.). 27 fictionalized essays by noted anthropologists examine religion, customs, government, additional facets of life among the Winnebago, Crow, Zuni, Eskimo, other tribes. 480pp. 6⅛ x 9¼. 27377-6 Pa. $10.95

FRANK LLOYD WRIGHT'S HOLLYHOCK HOUSE, Donald Hoffmann. Lavishly illustrated, carefully documented study of one of Wright's most controversial residential designs. Over 120 photographs, floor plans, elevations, etc. Detailed perceptive text by noted Wright scholar. Index. 128pp. 9¼ x 10¾. 27133-1 Pa. $11.95

THE MALE AND FEMALE FIGURE IN MOTION: 60 Classic Photographic Sequences, Eadweard Muybridge. 60 true-action photographs of men and women walking, running, climbing, bending, turning, etc., reproduced from rare 19th-century masterpiece. vi + 121pp. 9 x 12. 24745-7 Pa. $10.95

1001 QUESTIONS ANSWERED ABOUT THE SEASHORE, N. J. Berrill and Jacquelyn Berrill. Queries answered about dolphins, sea snails, sponges, starfish, fishes, shore birds, many others. Covers appearance, breeding, growth, feeding, much more. 305pp. 5¼ x 8¼. 23366-9 Pa. $8.95

GUIDE TO OWL WATCHING IN NORTH AMERICA, Donald S. Heintzelman. Superb guide offers complete data and descriptions of 19 species: barn owl, screech owl, snowy owl, many more. Expert coverage of owl-watching equipment, conservation, migrations and invasions, etc. Guide to observing sites. 84 illustrations. xiii + 193pp. 5⅜ x 8½. 27344-X Pa. $8.95

MEDICINAL AND OTHER USES OF NORTH AMERICAN PLANTS: A Historical Survey with Special Reference to the Eastern Indian Tribes, Charlotte Erichsen-Brown. Chronological historical citations document 500 years of usage of plants, trees, shrubs native to eastern Canada, northeastern U.S. Also complete identifying information. 343 illustrations. 544pp. 6½ x 9¼. 25951-X Pa. $12.95

STORYBOOK MAZES, Dave Phillips. 23 stories and mazes on two-page spreads: Wizard of Oz, Treasure Island, Robin Hood, etc. Solutions. 64pp. 8¼ x 11. 23628-5 Pa. $2.95

NEGRO FOLK MUSIC, U.S.A., Harold Courlander. Noted folklorist's scholarly yet readable analysis of rich and varied musical tradition. Includes authentic versions of over 40 folk songs. Valuable bibliography and discography. xi + 324pp. 5⅜ x 8½. 27350-4 Pa. $9.95

MOVIE-STAR PORTRAITS OF THE FORTIES, John Kobal (ed.). 163 glamor, studio photos of 106 stars of the 1940s: Rita Hayworth, Ava Gardner, Marlon Brando, Clark Gable, many more. 176pp. 8⅜ x 11¼. 23546-7 Pa. $12.95

BENCHLEY LOST AND FOUND, Robert Benchley. Finest humor from early 30s, about pet peeves, child psychologists, post office and others. Mostly unavailable elsewhere. 73 illustrations by Peter Arno and others. 183pp. 5⅜ x 8½. 22410-4 Pa. $6.95

YEKL and THE IMPORTED BRIDEGROOM AND OTHER STORIES OF YIDDISH NEW YORK, Abraham Cahan. Film Hester Street based on Yekl (1896). Novel, other stories among first about Jewish immigrants on N.Y.'s East Side. 240pp. 5⅜ x 8½. 22427-9 Pa. $6.95

SELECTED POEMS, Walt Whitman. Generous sampling from *Leaves of Grass*. Twenty-four poems include "I Hear America Singing," "Song of the Open Road," "I Sing the Body Electric," "When Lilacs Last in the Dooryard Bloom'd," "O Captain! My Captain!"–all reprinted from an authoritative edition. Lists of titles and first lines. 128pp. 5³⁄₁₆ x 8¼. 26878-0 Pa. $1.00

THE BEST TALES OF HOFFMANN, E. T. A. Hoffmann. 10 of Hoffmann's most important stories: "Nutcracker and the King of Mice," "The Golden Flowerpot," etc. 458pp. 5⅜ x 8½. 21793-0 Pa. $9.95

FROM FETISH TO GOD IN ANCIENT EGYPT, E. A. Wallis Budge. Rich detailed survey of Egyptian conception of "God" and gods, magic, cult of animals, Osiris, more. Also, superb English translations of hymns and legends. 240 illustrations. 545pp. 5⅜ x 8½. 25803-3 Pa. $13.95

FRENCH STORIES/CONTES FRANÇAIS: A Dual-Language Book, Wallace Fowlie. Ten stories by French masters, Voltaire to Camus: "Micromegas" by Voltaire; "The Atheist's Mass" by Balzac; "Minuet" by de Maupassant; "The Guest" by Camus, six more. Excellent English translations on facing pages. Also French-English vocabulary list, exercises, more. 352pp. 5⅜ x 8½. 26443-2 Pa. $8.95

CHICAGO AT THE TURN OF THE CENTURY IN PHOTOGRAPHS: 122 Historic Views from the Collections of the Chicago Historical Society, Larry A. Viskochil. Rare large-format prints offer detailed views of City Hall, State Street, the Loop, Hull House, Union Station, many other landmarks, circa 1904-1913. Introduction. Captions. Maps. 144pp. 9⅜ x 12¼. 24656-6 Pa. $12.95

OLD BROOKLYN IN EARLY PHOTOGRAPHS, 1865-1929, William Lee Younger. Luna Park, Gravesend race track, construction of Grand Army Plaza, moving of Hotel Brighton, etc. 157 previously unpublished photographs. 165pp. 8⅞ x 11¾. 23587-4 Pa. $13.95

THE MYTHS OF THE NORTH AMERICAN INDIANS, Lewis Spence. Rich anthology of the myths and legends of the Algonquins, Iroquois, Pawnees and Sioux, prefaced by an extensive historical and ethnological commentary. 36 illustrations. 480pp. 5⅜ x 8½. 25967-6 Pa. $8.95

AN ENCYCLOPEDIA OF BATTLES: Accounts of Over 1,560 Battles from 1479 B.C. to the Present, David Eggenberger. Essential details of every major battle in recorded history from the first battle of Megiddo in 1479 B.C. to Grenada in 1984. List of Battle Maps. New Appendix covering the years 1967-1984. Index. 99 illustrations. 544pp. 6½ x 9¼. 24913-1 Pa. $14.95

SAILING ALONE AROUND THE WORLD, Captain Joshua Slocum. First man to sail around the world, alone, in small boat. One of great feats of seamanship told in delightful manner. 67 illustrations. 294pp. 5⅜ x 8½. 20326-3 Pa. $5.95

ANARCHISM AND OTHER ESSAYS, Emma Goldman. Powerful, penetrating, prophetic essays on direct action, role of minorities, prison reform, puritan hypocrisy, violence, etc. 271pp. 5⅜ x 8½. 22484-8 Pa. $6.95

MYTHS OF THE HINDUS AND BUDDHISTS, Ananda K. Coomaraswamy and Sister Nivedita. Great stories of the epics; deeds of Krishna, Shiva, taken from puranas, Vedas, folk tales; etc. 32 illustrations. 400pp. 5⅜ x 8½. 21759-0 Pa. $10.95

BEYOND PSYCHOLOGY, Otto Rank. Fear of death, desire of immortality, nature of sexuality, social organization, creativity, according to Rankian system. 291pp. 5⅜ x 8½. 20485-5 Pa. $8.95

A THEOLOGICO-POLITICAL TREATISE, Benedict Spinoza. Also contains unfinished Political Treatise. Great classic on religious liberty, theory of government on common consent. R. Elwes translation. Total of 421pp. 5⅜ x 8½. 20249-6 Pa. $9.95

MY BONDAGE AND MY FREEDOM, Frederick Douglass. Born a slave, Douglass became outspoken force in antislavery movement. The best of Douglass' autobiographies. Graphic description of slave life. 464pp. 5⅜ x 8½. 22457-0 Pa. $8.95

FOLLOWING THE EQUATOR: A Journey Around the World, Mark Twain. Fascinating humorous account of 1897 voyage to Hawaii, Australia, India, New Zealand, etc. Ironic, bemused reports on peoples, customs, climate, flora and fauna, politics, much more. 197 illustrations. 720pp. 5⅜ x 8½. 26113-1 Pa. $15.95

THE PEOPLE CALLED SHAKERS, Edward D. Andrews. Definitive study of Shakers: origins, beliefs, practices, dances, social organization, furniture and crafts, etc. 33 illustrations. 351pp. 5⅜ x 8½. 21081-2 Pa. $8.95

THE MYTHS OF GREECE AND ROME, H. A. Guerber. A classic of mythology, generously illustrated, long prized for its simple, graphic, accurate retelling of the principal myths of Greece and Rome, and for its commentary on their origins and significance. With 64 illustrations by Michelangelo, Raphael, Titian, Rubens, Canova, Bernini and others. 480pp. 5⅜ x 8½. 27584-1 Pa. $9.95

PSYCHOLOGY OF MUSIC, Carl E. Seashore. Classic work discusses music as a medium from psychological viewpoint. Clear treatment of physical acoustics, auditory apparatus, sound perception, development of musical skills, nature of musical feeling, host of other topics. 88 figures. 408pp. 5⅜ x 8½. 21851-1 Pa. $10.95

THE PHILOSOPHY OF HISTORY, Georg W. Hegel. Great classic of Western thought develops concept that history is not chance but rational process, the evolution of freedom. 457pp. 5⅜ x 8½. 20112-0 Pa. $9.95

THE BOOK OF TEA, Kakuzo Okakura. Minor classic of the Orient: entertaining, charming explanation, interpretation of traditional Japanese culture in terms of tea ceremony. 94pp. 5⅜ x 8½. 20070-1 Pa. $3.95

LIFE IN ANCIENT EGYPT, Adolf Erman. Fullest, most thorough, detailed older account with much not in more recent books, domestic life, religion, magic, medicine, commerce, much more. Many illustrations reproduce tomb paintings, carvings, hieroglyphs, etc. 597pp. 5⅜ x 8½. 22632-8 Pa. $11.95

SUNDIALS, Their Theory and Construction, Albert Waugh. Far and away the best, most thorough coverage of ideas, mathematics concerned, types, construction, adjusting anywhere. Simple, nontechnical treatment allows even children to build several of these dials. Over 100 illustrations. 230pp. 5⅜ x 8½. 22947-5 Pa. $7.95

DYNAMICS OF FLUIDS IN POROUS MEDIA, Jacob Bear. For advanced students of ground water hydrology, soil mechanics and physics, drainage and irrigation engineering, and more. 335 illustrations. Exercises, with answers. 784pp. 6⅛ x 9¼.
65675-6 Pa. $19.95

SONGS OF EXPERIENCE: Facsimile Reproduction with 26 Plates in Full Color, William Blake. 26 full-color plates from a rare 1826 edition. Includes "The Tyger," "London," "Holy Thursday," and other poems. Printed text of poems. 48pp. 5¼ x 7.
24636-1 Pa. $4.95

OLD-TIME VIGNETTES IN FULL COLOR, Carol Belanger Grafton (ed.). Over 390 charming, often sentimental illustrations, selected from archives of Victorian graphics—pretty women posing, children playing, food, flowers, kittens and puppies, smiling cherubs, birds and butterflies, much more. All copyright-free. 48pp. 9¼ x 12¼.
27269-9 Pa. $7.95

PERSPECTIVE FOR ARTISTS, Rex Vicat Cole. Depth, perspective of sky and sea, shadows, much more, not usually covered. 391 diagrams, 81 reproductions of drawings and paintings. 279pp. 5⅜ x 8½. 22487-2 Pa. $7.95

DRAWING THE LIVING FIGURE, Joseph Sheppard. Innovative approach to artistic anatomy focuses on specifics of surface anatomy, rather than muscles and bones. Over 170 drawings of live models in front, back and side views, and in widely varying poses. Accompanying diagrams. 177 illustrations. Introduction. Index. 144pp. 8⅜ x11¼. 26723-7 Pa. $8.95

GOTHIC AND OLD ENGLISH ALPHABETS: 100 Complete Fonts, Dan X. Solo. Add power, elegance to posters, signs, other graphics with 100 stunning copyright-free alphabets: Blackstone, Dolbey, Germania, 97 more–including many lower-case, numerals, punctuation marks. 104pp. 8⅛ x 11. 24695-7 Pa. $8.95

HOW TO DO BEADWORK, Mary White. Fundamental book on craft from simple projects to five-bead chains and woven works. 106 illustrations. 142pp. 5⅜ x 8. 20697-1 Pa. $4.95

THE BOOK OF WOOD CARVING, Charles Marshall Sayers. Finest book for beginners discusses fundamentals and offers 34 designs. "Absolutely first rate . . . well thought out and well executed."–E. J. Tangerman. 118pp. 7¾ x 10⅝. 23654-4 Pa. $6.95

ILLUSTRATED CATALOG OF CIVIL WAR MILITARY GOODS: Union Army Weapons, Insignia, Uniform Accessories, and Other Equipment, Schuyler, Hartley, and Graham. Rare, profusely illustrated 1846 catalog includes Union Army uniform and dress regulations, arms and ammunition, coats, insignia, flags, swords, rifles, etc. 226 illustrations. 160pp. 9 x 12. 24939-5 Pa. $10.95

WOMEN'S FASHIONS OF THE EARLY 1900s: An Unabridged Republication of "New York Fashions, 1909," National Cloak & Suit Co. Rare catalog of mail-order fashions documents women's and children's clothing styles shortly after the turn of the century. Captions offer full descriptions, prices. Invaluable resource for fashion, costume historians. Approximately 725 illustrations. 128pp. 8⅜ x 11¼. 27276-1 Pa. $11.95

THE 1912 AND 1915 GUSTAV STICKLEY FURNITURE CATALOGS, Gustav Stickley. With over 200 detailed illustrations and descriptions, these two catalogs are essential reading and reference materials and identification guides for Stickley furniture. Captions cite materials, dimensions and prices. 112pp. 6½ x 9¼. 26676-1 Pa. $9.95

EARLY AMERICAN LOCOMOTIVES, John H. White, Jr. Finest locomotive engravings from early 19th century: historical (1804–74), main-line (after 1870), special, foreign, etc. 147 plates. 142pp. 11⅜ x 8¼. 22772-3 Pa. $10.95

THE TALL SHIPS OF TODAY IN PHOTOGRAPHS, Frank O. Braynard. Lavishly illustrated tribute to nearly 100 majestic contemporary sailing vessels: Amerigo Vespucci, Clearwater, Constitution, Eagle, Mayflower, Sea Cloud, Victory, many more. Authoritative captions provide statistics, background on each ship. 190 black-and-white photographs and illustrations. Introduction. 128pp. 8⅞ x 11¾. 27163-3 Pa. $13.95

EARLY NINETEENTH-CENTURY CRAFTS AND TRADES, Peter Stockham (ed.). Extremely rare 1807 volume describes to youngsters the crafts and trades of the day: brickmaker, weaver, dressmaker, bookbinder, ropemaker, saddler, many more. Quaint prose, charming illustrations for each craft. 20 black-and-white line illustrations. 192pp. 4⅝ x 6. 27293-1 Pa. $4.95

VICTORIAN FASHIONS AND COSTUMES FROM HARPER'S BAZAR, 1867–1898, Stella Blum (ed.). Day costumes, evening wear, sports clothes, shoes, hats, other accessories in over 1,000 detailed engravings. 320pp. 9⅜ x 12¼. 22990-4 Pa. $14.95

GUSTAV STICKLEY, THE CRAFTSMAN, Mary Ann Smith. Superb study surveys broad scope of Stickley's achievement, especially in architecture. Design philosophy, rise and fall of the Craftsman empire, descriptions and floor plans for many Craftsman houses, more. 86 black-and-white halftones. 31 line illustrations. Introduction 208pp. 6½ x 9¼. 27210-9 Pa. $9.95

THE LONG ISLAND RAIL ROAD IN EARLY PHOTOGRAPHS, Ron Ziel. Over 220 rare photos, informative text document origin (1844) and development of rail service on Long Island. Vintage views of early trains, locomotives, stations, passengers, crews, much more. Captions. 8⅞ x 11¾. 26301-0 Pa. $13.95

THE BOOK OF OLD SHIPS: From Egyptian Galleys to Clipper Ships, Henry B. Culver. Superb, authoritative history of sailing vessels, with 80 magnificent line illustrations. Galley, bark, caravel, longship, whaler, many more. Detailed, informative text on each vessel by noted naval historian. Introduction. 256pp. 5⅜ x 8½. 27332-6 Pa. $7.95

TEN BOOKS ON ARCHITECTURE, Vitruvius. The most important book ever written on architecture. Early Roman aesthetics, technology, classical orders, site selection, all other aspects. Morgan translation. 331pp. 5⅜ x 8½. 20645-9 Pa. $8.95

THE HUMAN FIGURE IN MOTION, Eadweard Muybridge. More than 4,500 stopped-action photos, in action series, showing undraped men, women, children jumping, lying down, throwing, sitting, wrestling, carrying, etc. 390pp. 7⅞ x 10⅝. 20204-6 Clothbd. $25.95

TREES OF THE EASTERN AND CENTRAL UNITED STATES AND CANADA, William M. Harlow. Best one-volume guide to 140 trees. Full descriptions, woodlore, range, etc. Over 600 illustrations. Handy size. 288pp. 4½ x 6⅜. 20395-6 Pa. $6.95

SONGS OF WESTERN BIRDS, Dr. Donald J. Borror. Complete song and call repertoire of 60 western species, including flycatchers, juncoes, cactus wrens, many more–includes fully illustrated booklet. Cassette and manual 99913-0 $8.95

GROWING AND USING HERBS AND SPICES, Milo Miloradovich. Versatile handbook provides all the information needed for cultivation and use of all the herbs and spices available in North America. 4 illustrations. Index. Glossary. 236pp. 5⅜ x 8½. 25058-X Pa. $6.95

BIG BOOK OF MAZES AND LABYRINTHS, Walter Shepherd. 50 mazes and labyrinths in all–classical, solid, ripple, and more–in one great volume. Perfect inexpensive puzzler for clever youngsters. Full solutions. 112pp. 8⅛ x 11. 22951-3 Pa. $4.95

PIANO TUNING, J. Cree Fischer. Clearest, best book for beginner, amateur. Simple repairs, raising dropped notes, tuning by easy method of flattened fifths. No previous skills needed. 4 illustrations. 201pp. 5⅜ x 8½. 23267-0 Pa. $6.95

A SOURCE BOOK IN THEATRICAL HISTORY, A. M. Nagler. Contemporary observers on acting, directing, make-up, costuming, stage props, machinery, scene design, from Ancient Greece to Chekhov. 611pp. 5⅜ x 8½. 20515-0 Pa. $12.95

THE COMPLETE NONSENSE OF EDWARD LEAR, Edward Lear. All nonsense limericks, zany alphabets, Owl and Pussycat, songs, nonsense botany, etc., illustrated by Lear. Total of 320pp. 5⅜ x 8½. (USO) 20167-8 Pa. $6.95

VICTORIAN PARLOUR POETRY: An Annotated Anthology, Michael R. Turner. 117 gems by Longfellow, Tennyson, Browning, many lesser-known poets. "The Village Blacksmith," "Curfew Must Not Ring Tonight," "Only a Baby Small," dozens more, often difficult to find elsewhere. Index of poets, titles, first lines. xxiii + 325pp. 5⅝ x 8¼. 27044-0 Pa. $8.95

DUBLINERS, James Joyce. Fifteen stories offer vivid, tightly focused observations of the lives of Dublin's poorer classes. At least one, "The Dead," is considered a masterpiece. Reprinted complete and unabridged from standard edition. 160pp. 5$\frac{3}{16}$ x 8¼. 26870-5 Pa. $1.00

THE HAUNTED MONASTERY and THE CHINESE MAZE MURDERS, Robert van Gulik. Two full novels by van Gulik, set in 7th-century China, continue adventures of Judge Dee and his companions. An evil Taoist monastery, seemingly supernatural events; overgrown topiary maze hides strange crimes. 27 illustrations. 328pp. 5⅜ x 8½. 23502-5 Pa. $8.95

THE BOOK OF THE SACRED MAGIC OF ABRAMELIN THE MAGE, translated by S. MacGregor Mathers. Medieval manuscript of ceremonial magic. Basic document in Aleister Crowley, Golden Dawn groups. 268pp. 5⅜ x 8½. 23211-5 Pa. $8.95

NEW RUSSIAN-ENGLISH AND ENGLISH-RUSSIAN DICTIONARY, M. A. O'Brien. This is a remarkably handy Russian dictionary, containing a surprising amount of information, including over 70,000 entries. 366pp. 4½ x 6⅛. 20208-9 Pa. $9.95

HISTORIC HOMES OF THE AMERICAN PRESIDENTS, Second, Revised Edition, Irvin Haas. A traveler's guide to American Presidential homes, most open to the public, depicting and describing homes occupied by every American President from George Washington to George Bush. With visiting hours, admission charges, travel routes. 175 photographs. Index. 160pp. 8¼ x 11. 26751-2 Pa. $11.95

NEW YORK IN THE FORTIES, Andreas Feininger. 162 brilliant photographs by the well-known photographer, formerly with *Life* magazine. Commuters, shoppers, Times Square at night, much else from city at its peak. Captions by John von Hartz. 181pp. 9¼ x 10¾. 23585-8 Pa. $12.95

INDIAN SIGN LANGUAGE, William Tomkins. Over 525 signs developed by Sioux and other tribes. Written instructions and diagrams. Also 290 pictographs. 111pp. 6⅛ x 9¼. 22029-X Pa. $3.95

ANATOMY: A Complete Guide for Artists, Joseph Sheppard. A master of figure drawing shows artists how to render human anatomy convincingly. Over 460 illustrations. 224pp. 8⅜ x 11¼. 27279-6 Pa. $10.95

MEDIEVAL CALLIGRAPHY: Its History and Technique, Marc Drogin. Spirited history, comprehensive instruction manual covers 13 styles (ca. 4th century thru 15th). Excellent photographs; directions for duplicating medieval techniques with modern tools. 224pp. 8⅜ x 11¼. 26142-5 Pa. $12.95

DRIED FLOWERS: How to Prepare Them, Sarah Whitlock and Martha Rankin. Complete instructions on how to use silica gel, meal and borax, perlite aggregate, sand and borax, glycerine and water to create attractive permanent flower arrangements. 12 illustrations. 32pp. 5⅜ x 8½. 21802-3 Pa. $1.00

EASY-TO-MAKE BIRD FEEDERS FOR WOODWORKERS, Scott D. Campbell. Detailed, simple-to-use guide for designing, constructing, caring for and using feeders. Text, illustrations for 12 classic and contemporary designs. 96pp. 5⅜ x 8½. 25847-5 Pa. $2.95

SCOTTISH WONDER TALES FROM MYTH AND LEGEND, Donald A. Mackenzie. 16 lively tales tell of giants rumbling down mountainsides, of a magic wand that turns stone pillars into warriors, of gods and goddesses, evil hags, powerful forces and more. 240pp. 5⅜ x 8½. 29677-6 Pa. $6.95

THE HISTORY OF UNDERCLOTHES, C. Willett Cunnington and Phyllis Cunnington. Fascinating, well-documented survey covering six centuries of English undergarments, enhanced with over 100 illustrations: 12th-century laced-up bodice, footed long drawers (1795), 19th-century bustles, l9th-century corsets for men, Victorian "bust improvers," much more. 272pp. 5⅜ x 8¼. 27124-2 Pa. $9.95

ARTS AND CRAFTS FURNITURE: The Complete Brooks Catalog of 1912, Brooks Manufacturing Co. Photos and detailed descriptions of more than 150 now very collectible furniture designs from the Arts and Crafts movement depict davenports, settees, buffets, desks, tables, chairs, bedsteads, dressers and more, all built of solid, quarter-sawed oak. Invaluable for students and enthusiasts of antiques, Americana and the decorative arts. 80pp. 6½ x 9¼. 27471-3 Pa. $8.95

HOW WE INVENTED THE AIRPLANE: An Illustrated History, Orville Wright. Fascinating firsthand account covers early experiments, construction of planes and motors, first flights, much more. Introduction and commentary by Fred C. Kelly. 76 photographs. 96pp. 8¼ x 11. 25662-6 Pa. $8.95

THE ARTS OF THE SAILOR: Knotting, Splicing and Ropework, Hervey Garrett Smith. Indispensable shipboard reference covers tools, basic knots and useful hitches; handsewing and canvas work, more. Over 100 illustrations. Delightful reading for sea lovers. 256pp. 5⅜ x 8½. 26440-8 Pa. $7.95

FRANK LLOYD WRIGHT'S FALLINGWATER: The House and Its History, Second, Revised Edition, Donald Hoffmann. A total revision—both in text and illustrations—of the standard document on Fallingwater, the boldest, most personal architectural statement of Wright's mature years, updated with valuable new material from the recently opened Frank Lloyd Wright Archives. "Fascinating"—*The New York Times*. 116 illustrations. 128pp. 9¼ x 10¾. 27430-6 Pa. $11.95

PHOTOGRAPHIC SKETCHBOOK OF THE CIVIL WAR, Alexander Gardner. 100 photos taken on field during the Civil War. Famous shots of Manassas Harper's Ferry, Lincoln, Richmond, slave pens, etc. 244pp. 10⅝ x 8¼. 22731-6 Pa. $9.95

FIVE ACRES AND INDEPENDENCE, Maurice G. Kains. Great back-to-the-land classic explains basics of self-sufficient farming. The one book to get. 95 illustrations. 397pp. 5⅜ x 8½. 20974-1 Pa. $7.95

SONGS OF EASTERN BIRDS, Dr. Donald J. Borror. Songs and calls of 60 species most common to eastern U.S.: warblers, woodpeckers, flycatchers, thrushes, larks, many more in high-quality recording. Cassette and manual 99912-2 $9.95

A MODERN HERBAL, Margaret Grieve. Much the fullest, most exact, most useful compilation of herbal material. Gigantic alphabetical encyclopedia, from aconite to zedoary, gives botanical information, medical properties, folklore, economic uses, much else. Indispensable to serious reader. 161 illustrations. 888pp. 6½ x 9¼. 2-vol. set. (USO) Vol. I: 22798-7 Pa. $9.95
Vol. II: 22799-5 Pa. $9.95

HIDDEN TREASURE MAZE BOOK, Dave Phillips. Solve 34 challenging mazes accompanied by heroic tales of adventure. Evil dragons, people-eating plants, blood-thirsty giants, many more dangerous adversaries lurk at every twist and turn. 34 mazes, stories, solutions. 48pp. 8¼ x 11. 24566-7 Pa. $2.95

LETTERS OF W. A. MOZART, Wolfgang A. Mozart. Remarkable letters show bawdy wit, humor, imagination, musical insights, contemporary musical world; includes some letters from Leopold Mozart. 276pp. 5⅜ x 8½. 22859-2 Pa. $7.95

BASIC PRINCIPLES OF CLASSICAL BALLET, Agrippina Vaganova. Great Russian theoretician, teacher explains methods for teaching classical ballet. 118 illustrations. 175pp. 5⅜ x 8½. 22036-2 Pa. $5.95

THE JUMPING FROG, Mark Twain. Revenge edition. The original story of The Celebrated Jumping Frog of Calaveras County, a hapless French translation, and Twain's hilarious "retranslation" from the French. 12 illustrations. 66pp. 5⅜ x 8½. 22686-7 Pa. $3.95

BEST REMEMBERED POEMS, Martin Gardner (ed.). The 126 poems in this superb collection of 19th- and 20th-century British and American verse range from Shelley's "To a Skylark" to the impassioned "Renascence" of Edna St. Vincent Millay and to Edward Lear's whimsical "The Owl and the Pussycat." 224pp. 5⅜ x 8½. 27165-X Pa. $4.95

COMPLETE SONNETS, William Shakespeare. Over 150 exquisite poems deal with love, friendship, the tyranny of time, beauty's evanescence, death and other themes in language of remarkable power, precision and beauty. Glossary of archaic terms. 80pp. 5 3/16 x 8¼. 26686-9 Pa. $1.00

BODIES IN A BOOKSHOP, R. T. Campbell. Challenging mystery of blackmail and murder with ingenious plot and superbly drawn characters. In the best tradition of British suspense fiction. 192pp. 5⅜ x 8½. 24720-1 Pa. $6.95

THE WIT AND HUMOR OF OSCAR WILDE, Alvin Redman (ed.). More than 1,000 ripostes, paradoxes, wisecracks: Work is the curse of the drinking classes; I can resist everything except temptation; etc. 258pp. 5⅜ x 8½. 20602-5 Pa. $5.95

SHAKESPEARE LEXICON AND QUOTATION DICTIONARY, Alexander Schmidt. Full definitions, locations, shades of meaning in every word in plays and poems. More than 50,000 exact quotations. 1,485pp. 6½ x 9¼. 2-vol. set.
Vol. 1: 22726-X Pa. $16.95
Vol. 2: 22727-8 Pa. $16.95

SELECTED POEMS, Emily Dickinson. Over 100 best-known, best-loved poems by one of America's foremost poets, reprinted from authoritative early editions. No comparable edition at this price. Index of first lines. 64pp. 5$\frac{3}{16}$ x 8¼.
26466-1 Pa. $1.00

CELEBRATED CASES OF JUDGE DEE (DEE GOONG AN), translated by Robert van Gulik. Authentic 18th-century Chinese detective novel; Dee and associates solve three interlocked cases. Led to van Gulik's own stories with same characters. Extensive introduction. 9 illustrations. 237pp. 5⅜ x 8½. 23337-5 Pa. $6.95

THE MALLEUS MALEFICARUM OF KRAMER AND SPRENGER, translated by Montague Summers. Full text of most important witchhunter's "bible," used by both Catholics and Protestants. 278pp. 6⅝ x 10. 22802-9 Pa. $12.95

SPANISH STORIES/CUENTOS ESPAÑOLES: A Dual-Language Book, Angel Flores (ed.). Unique format offers 13 great stories in Spanish by Cervantes, Borges, others. Faithful English translations on facing pages. 352pp. 5⅜ x 8½.
25399-6 Pa. $8.95

THE CHICAGO WORLD'S FAIR OF 1893: A Photographic Record, Stanley Appelbaum (ed.). 128 rare photos show 200 buildings, Beaux-Arts architecture, Midway, original Ferris Wheel, Edison's kinetoscope, more. Architectural emphasis; full text. 116pp. 8¼ x 11. 23990-X Pa. $9.95

OLD QUEENS, N.Y., IN EARLY PHOTOGRAPHS, Vincent F. Seyfried and William Asadorian. Over 160 rare photographs of Maspeth, Jamaica, Jackson Heights, and other areas. Vintage views of DeWitt Clinton mansion, 1939 World's Fair and more. Captions. 192pp. 8⅞ x 11. 26358-4 Pa. $12.95

CAPTURED BY THE INDIANS: 15 Firsthand Accounts, 1750-1870, Frederick Drimmer. Astounding true historical accounts of grisly torture, bloody conflicts, relentless pursuits, miraculous escapes and more, by people who lived to tell the tale. 384pp. 5⅜ x 8½. 24901-8 Pa. $8.95

THE WORLD'S GREAT SPEECHES, Lewis Copeland and Lawrence W. Lamm (eds.). Vast collection of 278 speeches of Greeks to 1970. Powerful and effective models; unique look at history. 842pp. 5⅜ x 8½. 20468-5 Pa. $14.95

THE BOOK OF THE SWORD, Sir Richard F. Burton. Great Victorian scholar/adventurer's eloquent, erudite history of the "queen of weapons"–from prehistory to early Roman Empire. Evolution and development of early swords, variations (sabre, broadsword, cutlass, scimitar, etc.), much more. 336pp. 6⅛ x 9¼.
25434-8 Pa. $9.95

AUTOBIOGRAPHY: The Story of My Experiments with Truth, Mohandas K. Gandhi. Boyhood, legal studies, purification, the growth of the Satyagraha (nonviolent protest) movement. Critical, inspiring work of the man responsible for the freedom of India. 480pp. 5⅜ x 8½. (USO) 24593-4 Pa. $8.95

CELTIC MYTHS AND LEGENDS, T. W. Rolleston. Masterful retelling of Irish and Welsh stories and tales. Cuchulain, King Arthur, Deirdre, the Grail, many more. First paperback edition. 58 full-page illustrations. 512pp. 5⅜ x 8½. 26507-2 Pa. $9.95

THE PRINCIPLES OF PSYCHOLOGY, William James. Famous long course complete, unabridged. Stream of thought, time perception, memory, experimental methods; great work decades ahead of its time. 94 figures. 1,391pp. 5⅜ x 8½. 2-vol. set.
Vol. I: 20381-6 Pa. $12.95
Vol. II: 20382-4 Pa. $12.95

THE WORLD AS WILL AND REPRESENTATION, Arthur Schopenhauer. Definitive English translation of Schopenhauer's life work, correcting more than 1,000 errors, omissions in earlier translations. Translated by E. F. J. Payne. Total of 1,269pp. 5⅜ x 8½. 2-vol. set.
Vol. 1: 21761-2 Pa. $11.95
Vol. 2: 21762-0 Pa. $12.95

MAGIC AND MYSTERY IN TIBET, Madame Alexandra David-Neel. Experiences among lamas, magicians, sages, sorcerers, Bonpa wizards. A true psychic discovery. 32 illustrations. 321pp. 5⅜ x 8½. (USO) 22682-4 Pa. $8.95

THE EGYPTIAN BOOK OF THE DEAD, E. A. Wallis Budge. Complete reproduction of Ani's papyrus, finest ever found. Full hieroglyphic text, interlinear transliteration, word-for-word translation, smooth translation. 533pp. 6½ x 9¼.
21866-X Pa. $10.95

MATHEMATICS FOR THE NONMATHEMATICIAN, Morris Kline. Detailed, college-level treatment of mathematics in cultural and historical context, with numerous exercises. Recommended Reading Lists. Tables. Numerous figures. 641pp. 5⅜ x 8½.
24823-2 Pa. $11.95

THEORY OF WING SECTIONS: Including a Summary of Airfoil Data, Ira H. Abbott and A. E. von Doenhoff. Concise compilation of subsonic aerodynamic characteristics of NACA wing sections, plus description of theory. 350pp. of tables. 693pp. 5⅜ x 8½. 60586-8 Pa. $14.95

THE RIME OF THE ANCIENT MARINER, Gustave Doré, S. T. Coleridge. Doré's finest work; 34 plates capture moods, subtleties of poem. Flawless full-size reproductions printed on facing pages with authoritative text of poem. "Beautiful. Simply beautiful."–*Publisher's Weekly*. 77pp. 9¼ x 12. 22305-1 Pa. $6.95

NORTH AMERICAN INDIAN DESIGNS FOR ARTISTS AND CRAFTSPEOPLE, Eva Wilson. Over 360 authentic copyright-free designs adapted from Navajo blankets, Hopi pottery, Sioux buffalo hides, more. Geometrics, symbolic figures, plant and animal motifs, etc. 128pp. 8⅜ x 11. (EUK) 25341-4 Pa. $8.95

SCULPTURE: Principles and Practice, Louis Slobodkin. Step-by-step approach to clay, plaster, metals, stone; classical and modern. 253 drawings, photos. 255pp. 8⅛ x 11.
22960-2 Pa. $11.95

THE INFLUENCE OF SEA POWER UPON HISTORY, 1660–1783, A. T. Mahan. Influential classic of naval history and tactics still used as text in war colleges. First paperback edition. 4 maps. 24 battle plans. 640pp. 5⅜ x 8½.   25509-3 Pa. $12.95

THE STORY OF THE TITANIC AS TOLD BY ITS SURVIVORS, Jack Winocour (ed.). What it was really like. Panic, despair, shocking inefficiency, and a little heroism. More thrilling than any fictional account. 26 illustrations. 320pp. 5⅜ x 8½.
20610-6 Pa. $8.95

FAIRY AND FOLK TALES OF THE IRISH PEASANTRY, William Butler Yeats (ed.). Treasury of 64 tales from the twilight world of Celtic myth and legend: "The Soul Cages," "The Kildare Pooka," "King O'Toole and his Goose," many more. Introduction and Notes by W. B. Yeats. 352pp. 5⅜ x 8½.   26941-8 Pa. $8.95

BUDDHIST MAHAYANA TEXTS, E. B. Cowell and Others (eds.). Superb, accurate translations of basic documents in Mahayana Buddhism, highly important in history of religions. The Buddha-karita of Asvaghosha, Larger Sukhavativyuha, more. 448pp. 5⅜ x 8½.   25552-2 Pa. $12.95

ONE TWO THREE . . . INFINITY: Facts and Speculations of Science, George Gamow. Great physicist's fascinating, readable overview of contemporary science: number theory, relativity, fourth dimension, entropy, genes, atomic structure, much more. 128 illustrations. Index. 352pp. 5⅜ x 8½.   25664-2 Pa. $8.95

ENGINEERING IN HISTORY, Richard Shelton Kirby, et al. Broad, nontechnical survey of history's major technological advances: birth of Greek science, industrial revolution, electricity and applied science, 20th-century automation, much more. 181 illustrations. ". . . excellent . . ."–*Isis.* Bibliography. vii + 530pp. 5⅜ x 8¼.
26412-2 Pa. $14.95

DALÍ ON MODERN ART: The Cuckolds of Antiquated Modern Art, Salvador Dalí. Influential painter skewers modern art and its practitioners. Outrageous evaluations of Picasso, Cézanne, Turner, more. 15 renderings of paintings discussed. 44 calligraphic decorations by Dalí. 96pp. 5⅜ x 8½. (USO)   29220-7 Pa. $4.95

ANTIQUE PLAYING CARDS: A Pictorial History, Henry René D'Allemagne. Over 900 elaborate, decorative images from rare playing cards (14th–20th centuries): Bacchus, death, dancing dogs, hunting scenes, royal coats of arms, players cheating, much more. 96pp. 9¼ x 12¼.   29265-7 Pa. $11.95

MAKING FURNITURE MASTERPIECES: 30 Projects with Measured Drawings, Franklin H. Gottshall. Step-by-step instructions, illustrations for constructing handsome, useful pieces, among them a Sheraton desk, Chippendale chair, Spanish desk, Queen Anne table and a William and Mary dressing mirror. 224pp. 8⅛ x 11¼.
29338-6 Pa. $13.95

THE FOSSIL BOOK: A Record of Prehistoric Life, Patricia V. Rich et al. Profusely illustrated definitive guide covers everything from single-celled organisms and dinosaurs to birds and mammals and the interplay between climate and man. Over 1,500 illustrations. 760pp. 7½ x 10⅛.   29371-8 Pa. $29.95

Prices subject to change without notice.

Available at your book dealer or write for free catalog to Dept. GI, Dover Publications, Inc., 31 East 2nd St., Mineola, N.Y. 11501. Dover publishes more than 500 books each year on science, elementary and advanced mathematics, biology, music, art, literary history, social sciences and other areas.